COMPLETE PLANS
FOR BUILDING
HORSE BARNS
BIG AND SMALL

COMPLETE PLANS

FOR BUILDING

HORSE BARNS

BIG AND SMALL

By
NANCY W. AMBROSIANO

and

MARY F. HARCOURT

Breakthrough
Publications, Inc.

For information address:

Breakthrough Publications
310 North Highland Avenue
Ossining, NY 10562

International Standard Book Number: 0-914327-28-3

Designed by Jacques Chazaud
Illustrations by Mace Bell, Isabel Guera,
Eddie Hartlove and Steven Prifti

Manufactured in MEXICO

Fourth Printing 1993

CONTENTS

Chapter 6 COSTS   75

Chapter 7 CONSTRUCTION   76

Chapter 8 BARN PLANS   81

ACKNOWLEDGMENTS

Putting this book together has been a delightful and challenging project, much like building your own dream barn. The overall ideas come easy, but the execution is the hard part. So many people have helped, from lending technical assistance, to bright ideas, to plain old moral support, that a complete list is impossible. For one thing, it would have to include a few score of horses who've benefited and/or sometimes suffered patiently as barn inhabitants. Here's a try at listing them:

Hector Alcalde, Takaro, Middleburg, Virginia; All Afshari, Photo Memos; John Ambrosiano; Dr. J.M. Bowen, Chesapeake, Virginia Polytechnic Institute; Penny and Lloyd Burger; David and Sharon Cooper, Aldie, Virginia; Dr. J.R. Gardner, Virginia Polytechnic Institute; Paul and Phyllis Hassell, Hassell Arabians, Reddick, Florida; Mr. and Mrs. Frank Hensley, Breaking Dawn Stables, Keswick, Virginia; Fritz, Karen and Sacha Himmelmayer, Kafri Farm, Orange, Virginia; Dr. Arden Huff, extension horse specialist, Virginia Extension Service; Gordon Hammond, Bradford, Maine; Carmen Johnson, Windfield Station, Nicasio, California; Helen and Jack Junkin, Old Mill Run Farm, Mason Neck, Virginia; Fred Kohler, Bittersweet Farm, Middleburg, Virginia; Karen Kresge, Connection Newspaper Group, Fairfax, Virginia; Doug Linton, Custom Barn Builders, Clifton, Virginia; Helen Makarov, Middleburg, Virginia; Midwest Plan Service; Dr. Bob Mowrey, extension horse specialist, North Carolina Extension Service; Tolli Nelson, Collie Blue Ranch, North Pole, Alaska; Jan Neuharth, Paper Chase Farm, Middleburg, Virginia; Dr. Judith Novicki, Jacksonville, North Carolina; Steve Prifti; McGann Saphire, Marion County, Florida extension agent; Julie Saxelby, Nicasio Valley Arabians, Nicasio, California; Joe Ann Scott; September Farms, Ocala, Florida; Sherry Shriver; Lalla Rook "Lassie" Tompkins, Briar Patch Farm, Micapony, Florida; Upperville Barns, Division of Northern Counties, Upperville, Virginia; Dr. H.E. White, Virginia Polytechnic Institute; Doug and Singie Williams, Chestnut Hill Farm, Spring Grove, Virginia; P.J. Williams Inc., Somerset, Virginia; Richard O. Zirk, agricultural engineer, Staunton, Virginia; and many others.

FOREWORD

Throughout time, horses and humans have enjoyed a special, useful relationship. Today, there is more interest in horses than ever before. Horses are kept for a variety of reasons, including business, sport and recreation. In nature, the horse is an outdoor, forage-eating athlete. In this age, people have to provide shelter, food, care and exercise for the horse. This can be as simple or as elaborate as one may choose, but a cost-effective program should be a major consideration. Horse owners invest considerable time and money in horses, land and buildings. Evaluation and careful planning should be considered in the process of establishing, expanding or renovating facilities.

Establishment of a major facility involves sizable investment and the result, good or bad, is more or less permanent. This book addresses the issues and planning process, and it provides guides to effective horse care and a range of possible facilities. It is an excellent reference for horse owners, prospective horse owners, professional builders, educators and engineers. It would also be a useful text for advanced horse-production classes or schools.

Over the years, I have visited hundreds of facilities in all parts of the country. Often, we are asked for advice about plans, locations, renovations and problems. We can never prevent all the possible problems or flaws from cropping up, but this text should go a long way toward the goal of helping owners and industry.

Early in my career, I was involved in developing one of the first modern-day horse building-and-plans publications. This was later revised, and it is still a useful document, though technology continues to develop. This book brings new management techniques well into focus, which are needed by the industry, and is an excellent guide. Still, the user must visit working facilities, develop a plan, seek advice and draw up a program specific to his or her needs and user intent.

Most owners do an excellent job. However, I generally deal with problems, and there are many, often needlessly repeated. For example, two of the most common involve inadequate drainage and poor ventilation. Many stables are dusty and many have inadequate lighting. Labor-saving techniques can also be built into the planning process. As for economics: Who ever heard of that in a horse operation? But horse farms should be run like any other business. Materials handling, feed, hay, bedding and manure, are all too often an afterthought.

Safety for the horse, the worker and the public must be built into the system. Often we are asked after the fact: How do you like my renovation or location or overall facility? Frequently, existing facilities should not be renovated or used. Just because a fence is there or a building exists is no reason that it is useful. Many farms also have no safe way to control or move animals by planned fencing. Many stables are situated so that if a horse gets loose or out of a stall, it has immediate access to a major highway.

Finally, aesthetics, green space, fly control and pasture or paddock management are all important, especially in promoting and marketing the horse.

With all this in mind, we should return to the basics. Why do we have horses? One of the major reasons is to have fun. We often let the process of horse ownership override the purpose of ownership. You need riding and use areas. This book is dedicated to help you ride and use your horses for fun and profit.

Dr. Arden N. Hutt
Professor, Animal Sciences Department
Virginia Tech
Blacksburg, VA 24061

INTRODUCTION

As a reader of this book, you either already own horses or plan on becoming more deeply involved with them. Whatever category you are in, welcome to the horse industry. We use the word "industry" because even if you plan on having horses only for pleasure, you are still part of a vast industry of people who derive their livelihood and recreation from a sport as old as human beings' first domestication of the horse and as young as the delighted child receiving her first pony at Christmas.

Whoever said that the outside of a horse was good for the inside of a man clearly knew the therapeutic value of working with this animal, even discounting time spent riding. A large part of horsemanship, of course, is the day-to-day care of the creature—a time-consuming task at best. But while the care and upkeep of a horse require much daily involvement, the rewards you receive as you survey the world from the top of your favorite mount more than make up for the effort of ownership.

And that brings us to the ultimate purpose of this book. We would like to share with you some ideas that have helped many people enjoy horse ownership. Horseowners all over the country shared their successes and failures, enabling us to put together a book on the best methods of keeping horses.

That's the key for good horsepeople—seeking out others and learning from their experiences. Not only will you get information to help you with your plans, but you'll likely run into some interesting characters and become quickly absorbed in the local horse community.

In talking to good horsemen and women from Alaska to Florida, we've found they have three things in common. First, they have a genuine concern and liking for the horse. Second, they believe that the horse has a job to do and should perform that job well with the correct care and training. Third, they have developed an eye that always looks for the smallest detail, from the amount of water the horse consumes that day, to the quality of its performance, to the overall look of the animal. They incorporate new ideas and experiences into daily encounters with their horses.

Whether you keep horses as a hobby, business or recreation, they can bring you lifelong pleasure. We hope our book will help you become an even more capable horseowner with many, many years of pleasure in the horse industry.

1 CHOOSING YOUR BARN

Getting started in the horse world takes a few key decisions, right from the start. First, how large an operation do you see yourself tackling? Will you have to do the job by yourself, or will you have qualified assistance? Whether you want a small or a large operation, you need to decide whether you want to be in the horse industry for pleasure or profit.

If you decide to go for profit in the beginning, your knowledge, management and facility level needs are probably going to be higher, simply because you need to invest money to make money. But never forget that good management can be used in place of a large capital investment with good results.

Obviously, the for-profit procedure requires a business plan. But even small operations take planning too. After all, if you are in it only for pleasure, your only reward or profit will be the enjoyment you and your family derive from the ownership of horses. No matter whether your plans are large or small, begin by asking yourself the following questions:

Am I Choosing the Right Neighborhood?

While a man's home may be his castle, living next to neighbors who abhor horses can bring on the worst of border wars. If you can choose property in an area where the natives support and share your pastime, do

so. The extra amount you might spend for land in a "horsey" area can save you money and stress in the long run. At least, make a selection in a semirural area.

Be sure to check the zoning regulations for the prospective piece of property. Zoning will tell you the type and number of animals you are permitted to keep on a specific piece of property. Check your community covenants, as well, for restrictions on such details as barn size and style, manure disposal and fencing requirements. How to check both zoning and covenants is discussed in detail in Chapter 3.

When purchasing land, look to adjacent areas for riding privileges. Many communities are requiring easements and riding right-of-ways that allow you plenty of room to ride without having to own vast acreage. Access to trails or land where the owner has given you permission to ride keeps you from getting land-locked and riding only in a confined arena.

Don't depend on your realtor to know such information. It's best to get it straight from the horse's mouth, in this case, the local government offices. Be sure to check for restrictions, too, on professional riding instruction on your property, on whether you can legally board horses there, and on any other factors that may limit your full use of the area for equestrian activities.

What is the Largest Number of Horses I Might Keep?

The number of horses you can keep may be restricted by zoning, the amount of labor available or your other responsibilities. It is a key question, though, if you hope to have a barn plan that will suit you for the longest time.

Will I Do All the Work Myself?

If you plan to move your horses from a full-care boarding stable to your own facility, but have no experience in the day-to-day care of owning a horse, get some! The effort and work of caring for your horses on a day-to-day basis can be enjoyable if you know what you are in for.

The average working horseowner spends at least 50 percent of "horse time" in barn chores. That's a lot of effort if you're not committed to doing the work. Before you get too far into the project, decide whom the majority of care will fall on—you, a family member, or hired workers. How many other responsibilities does the key worker have? And how can you make the most efficient barn arrangement so you'll waste the least amount of that person's time?

This is the area that can kill children's interest in horsemanship, so you have a couple of choices if you'd like to keep their enthusiasm. Parents can either do the work themselves or hire workers, which defeats the whole idea of children owning horses, or they can design a facility that is as labor-saving as possible and then have a firm family understanding about the responsibilities horse ownership entails.

If this operation will be organized like the traditional American family with all members pitching in, the work load becomes easier and more horses can be comfortably kept. However, if you are living alone and are unable to afford boarding at a full-care facility, then you must be prepared to assume all responsibilities for the upkeep of all the horses you plan on owning—a major and time-consuming responsibility depending on how you plan on maintaining the horse or horses you own.

What Are My Options in Caring for a Horse?

Depending on the climate and your preferences, horses can be kept simply or elaborately. The most simple method is totally on pasture. A good-quality pasture with adequate clean water can supply the total nutrient needs for an average pleasure horse during good weather conditions. If you add a salt block and some protection from sun or wind, a thrifty pleasure horse can be comfortable with little effort on your part. Then, too, your horse is far less likely to develop bad habits from boredom or respiratory trouble from an enclosed environment.

During cold or hot, dry weather when grass may be sparse, you may have to supplement a pasture with hay and grain. And if your local weather varies little in temperature, wind, sun and precipitation, it's possible you can get by with trees for natural shelter or a simple run-in shed. The horseperson must be aware of times when drought restricts water or grass supply and supplement accordingly, but generally, if your fences are safe, your grass and water supply good, and you check closely and regularly on your animal, pasture is the easiest method of keeping a horse.

If you are actively participating in serious competition, however, you may feel more comfortable having a stable. It gives you more control over the horse's environment, allowing such things as clipping your horse's winter coat, housing him inside to reduce summer sun bleaching, and keeping a closer eye on possible veterinary problems.

Thus, you're facing a big choice; Should you opt for the simplistic run-in shed or the more traditional barn/stall arrangement? Here is where dreaming can be both fun and profitable. By dreaming about what you want, you can begin to decide what you actually need and can afford. You can get a good start by examining existing facilities. It helps if you have worked in or around enough facilities to have a real feel for your preferences. Then you can get down to the actual project secure in the knowledge that you have done your homework and that the facility you build will be perfect.

Of course, keep in mind that nothing is ever totally perfect. There is always a better mousetrap or idea just around the corner; what you build today may not always suit all your needs. But if you plan carefully, you can develop a flexible enough arrangement to cover most reasonable changes. So get a firm idea of what your needs are, develop a plan and be willing to be flexible. Then get started. That's probably the hardest part.

2 ADVICE AND INFORMATION SOURCES

SOURCES OF HELP

Now that you've made the decision to build your own facility, no doubt you have plenty of questions about your upcoming effort and a vague feeling that there's more you have yet to think of.

There's only one way to find out—ask questions. Help comes in three varieties: Free help that is worth something, free help that is best ignored, and valuable help you pay for.

After a while you'll be able to identify the free help that is useless. There are lots of "experts" in the horse industry who are only too glad to tell everything they know—but that may get you in more trouble than it's worth. Conversely, there are an enormous number of knowledgeable horsepeople who are more than willing to offer good advice and assistance. Time and experience will help you tell the difference.

Many horse-related magazines offer excellent articles on horse management and facilities. Subscribing to one or two of the better ones will give you some fine ideas for the day-to-day working of your barn and management of your horses.

There are many professional horsepeople who have moved into the world of consulting and can help you with everything from selection of the right property or horse to international marketing, and everything in between. Farriers, veterinarians and feed dealers can be great sources of help. Another excellent source of good, free information are the local county government offices serving the agricultural community.

One of these, the Cooperative Extension Service (CES), has a local office in nearly every county in the nation.

Originally started in the early part of this century as an extension of the land-grant universities, CES's purpose was to improve the quality of life for those living in rural environments. Today, its purpose is to disseminate information, generally free of charge (or at nominal fees), on a wide variety of agriculturally related subjects, many of which involve keeping and maintaining horses. This information is available to everyone, regardless of where you live or how complex your plans are.

CES staff members, called "county agents," "extension agents" or "agricultural, youth and/or home economic agents" (depending on your state) can not only give you first-hand information on pastures, facilities, feeds, feeding, fencing, and a host of other topics, but they also have access to the knowledge bank at the state agricultural college. Many extension offices have expanded their services to include information on farm planning, record-keeping and computer programs for farms. All in all, the Cooperative Extension Office may have so much information to share with you that it may take you several visits to receive the full benefit of its services.

A federal office whose job is related to land management and whose office is often located near or in the same area as the Cooperative Extension Service is the Soil Conservation Service (SCS). It is a service of the U.S. Department of Agriculture and has offices in every state, with staff members who have extensive knowledge about soil and soil usage in their jurisdictions.

Knowing if the land you are purchasing (or already own) is suitable for livestock/pasture, the best location for your house, your roads, your barns and so forth is valuable information. The SCS office can offer you assistance in identifying the type of soil in your area, its best uses according to its profile, and how you can overcome some of the not-so-desirable parcels you may be forced to deal with. Consulting SCS can save you money from the start—forewarned is fore armed.

Finally, another government agency that you may or may not need, depending on the type (pleasure or profit) and scope of your operation, is the Agricultural Stabilization and Conservation Service (ASCS). Again, you can often find it near other agriculturally related services. Its function is to regulate the dissemination of government funds for the agricultural industry in this country. It has funding for assistance in establishing/maintaining certain crops/pasture lands in given regions, and your farm may be eligible if it meets the ASCS criteria.

Don't overlook your county or city planning department and other local government offices that are concerned with land use. They often have staff members familiar with pertinent subjects. Local community colleges, if they have agricultural programs,

are also a fine source of help, as is the Federal Department of Parks and Recreation in your area and the National Forestry Service.

A word about government bureaucracy: The Cooperative Extension Service and the SCS office have few or no strings attached to their advice, and few forms. The ASCS offices are more tightly regulated and require more paper work, as money is often involved in their transactions. You'll have to decide if the rewards (monies received) are worth the effort in dealing with this office. Ask questions. You, as an ordinary horse owner, may not qualify for certain things, but there are some funds available on a limited basis that can be of help to your operation.

Of course, the key to sorting through all the answers you get is organization. By organizing the information you've gathered as to type and source you'll eventually have a file that can help you put together a coherent picture.

Set up your files according to the categories of information you need to build your facility. Some suggestions include, but are not limited to:

- Fencing
- Barn plans
- Construction materials
- Ventilation
- Plumbing and water sources
- Electrical wiring/lighting
- Arrangement of inner-barn facilities (feed rooms, wash pits, tack rooms and so on)
- Farm layout plans
- Government (federal, state, local) offices offering help and contact people
- Same list for regulations
- Reliable sources of information such as knowledgeable horsepeople, governmental agencies and consultants

WORKING WITH BUILDERS

If you see the construction of a full-sized barn as too time consuming a task, there are plenty of professionals and talented amateurs ready to take over for you. However, before you turn your land and your checkbook over to them, be sure you all see eye to eye on what's being built.

A word of caution if you're also building a house: The people who build your new house, unless they

are also farm folks, may not necessarily be your first choice as barn builders, merely because you are asking them to build a structure that's outside their area of expertise. Basic as it is, a barn has key requirements of strength, drainage and ventilation that the inexperienced will have to think twice about.

If there is a barn builder in the area, look into his rates either as first in charge on the site or at least as a consultant. For example, he can ensure that boards are always nailed to the horse's side of the post, that doors swing in the right direction, (but preferably slide), and that all the other minutiae peculiar to barns are done right.

The barn builder will generally give you the best price, because he knows exactly what his local costs are. A home builder, used to estimating such things as drywall and concrete slabs, may be dealing with too complex an equation for a "simple" barn.

If you aren't up for the job yourself and there are no barn builders available, then take your home builder on a tour of some well-done barns. Impress upon him that the structure must be horse-proof—a horse's needs must come first in a barn. Overestimating is as useless as underestimating, of course, so be prepared to give your builder a realistic picture of life around a barn.

3 THE LAND AND THE LAW

ZONING, DEEDS AND COVENANTS

Before you buy a single board for your barn, you must be prepared to conform to myriads of local requirements, even though what you're building is "only a barn." If you thought zoning requirements are only for houses and factories, you'll need to bone up on the local view of agriculture and recreational land use. Zoning and building codes are governed by the state and also by the local county or city in which you build your barn.

Some communities may not have any requirements beyond those established by the state for land use and building safety, which are part of an effort to ensure the quality of life of their communities. Other regions have added strict requirements to the basic zoning and building codes. The key is to know what your area allows or demands, specifically with regard to keeping livestock and the buildings that go along with it. Before you start, find out what permits are called for before and during construction. Plumbing, electrical work, footings, drainage, all come under some inspector's jurisdiction, and you ignore them at your peril.

Most states and counties divide up land according to its preferred use. General classifications are:

- Agricultural
- Residential
- Commercial or business
- Industrial

These categories can be further divided so that what land you own may be tightly restricted as to the number and kind of animals allowed, what they're used for, where on the property their housing is placed, how it's fenced, and much more. The local zoning office can tell you exactly what you can and cannot do with your land.

Never take anyone's word about the zoning on a piece of land, and don't make assumptions about its zoning by looking at what surrounds it. Always go to the main source, the zoning commission.

In addition to regional limitations on horsekeeping, the community may have its own covenants—restrictions designed to keep a consistent appearance in particular neighborhoods. Also, the person or land company selling the property may place some restrictions in the deed. Read this carefully and consult a lawyer if you are in doubt on any sections. By protecting the neighborhood you are protecting yourself from others moving in and running undesirable operations too close to you.

Covenants, deed restrictions and even zoning can sometimes be waived if the reasons are compelling enough, but the bureaucratic tangle can be daunting. Ask yourself if the land is truly worth the trouble before you start, as waiver negotiations can slow you down considerably.

Once you are aware of the restrictions on your property, you can plan a farm design that complies and begin filing for building permits. Here again, the range in regulations extends from no permit required for agricultural buildings in areas zoned for them to barns needing the same code inspections as those for a regular suburban home.

Call or visit the building inspector's office to see about "blanket" permits that give you permission to go on with more than one small project at a time. While the aggravation of dealing with all these regulations for a simple, uninhabited structure may seem too much, be persistent.

If you work with, instead of against, the inspectors and planners, your entire project will be simpler. In fact, some barn builders, anticipating problems with inspectors who want house-type construction used in every case, have had good luck when they've made the inspector almost an advisor on the whole project. This way, they benefit from the inspector's local experience on land, materials, workers, etc., and they avoid an adversarial relationship that can hold up the whole project. Be especially attentive to this if you plan a house/barn combination or a barn with many extras, such as lounges and bathrooms.

ACREAGE

You will want to have pastures for your animals, if at all possible. Horses, by nature, are grazers, and will do better if they have the opportunity to move and eat constantly as is their natural inclination. This both provides them with daily exercise and reduces your feed costs.

In certain areas, due to dense population, high prices and lack of available ground for pastures, horses are maintained very successfully on small plots. This type of horsekeeping requires good horse-management practices, as close confinement can increase some problems such as high worm counts or sand colic from eating off bare ground.

Erosion and water runoff are problems with these small dirt paddocks unless their location is chosen carefully. Use a small strip of grass as a buffer around the paddock to help prevent rapid water runoff. Consider applying mulch to high-traffic areas, such as around fence lines, to reduce dust.

Since the conditions for pastures differ not only from state to state but from county to county, it's important that you check with local experts to get the best information on your soil and climate type. Help can be found from the Cooperative Extension Service or from private farm consultants who can help you map a pasture plan that is specifically geared to your situation and meets your needs.

The American Horse Council, in Washington, D.C. has recently published an excellent guideline for zoning and urban horsekeeping. In it, they point out that the amount of space is not so important as is the way the space is maintained. It must be safely and well fenced and maintained with special care to manage dust, noise, manure, odor, flies and especially rodents.

Animals Per Acre

The number of animals recommended per acre of land is a variable that can fall into such a wide range that it's best to consult local people. In areas where the climate is warm year round and the rainfall ample, you may actually be able to figure one horse per acre and expect it to get most of its nutrients from pasture. The other extreme is a climate which is so dry and sparse that it takes five or more acres per animal to provide any kind of nutrients.

Generally, if you live in a seasonal climate area and plan on feeding a horse hay and grain in addition to pasture, estimate two acres per horse. During warm months, you may be able to cut supplemental feeding to nothing, depending on how much you're using your horse and how easy he or she is to keep.

Variety of Grass

Through experimentation, the agricultural industry has developed a large variety of grasses that can be targeted to withstand drought, high rainfall, dense soils, close grazing and cool as well as warm seasons, so getting expert help can enable you to plan the right variety for your specific situation.

Not just any old grass will do. Grass requires judicious care and constant upgrading. Horses are highly destructive to pastures, and, in some cases, the pasture can fight back. For example, many common types of grass known as fescues carry a fungus that causes abortion in broodmares. If you buy land for a breeding operation and have acres of this grass, you're in for complete reseeding if you want to eliminate worry. Other grass types, while harmless to horses, are just not strong enough to take the pressure of constant grazing. They go from green to extinction in one season.

If you are establishing a pasture for the first time, the most difficult thing to do after seeding and seeing the green grass grow is to wait until your pasture matures. General recommendations are to seed in the spring or fall and to keep livestock off the pasture until it is truly well established—not just looking a little greenish. This may take a full year or even two, depending on the climate and type of grass you plant.

Failure to wait until your grass has put down a strong root system will result in your horses pulling the new tender grass shoots up, roots and all, thereby leaving nothing for new growth. Good pastures, having an established root system with a reserve of nutrients to allow refoliation, enable the leaves to replace themselves as they are grazed. As a rule, horses should be removed from a pasture when plant height is reduced to one or two inches.

Stretching Your Pasture

One method of maximizing the use of your pasture is to plant both warm and cool season perennials or annuals. If your varieties are compatible, you can use the property year round.

The best method for stretching pastures is management. The following four management skills and techniques will add life and quality to your pasture.

1. Rotate animals. If at all possible, rotate your animals on the available pasture land. Sectioning off the pasture is a cheap way to force your animals to completely graze an area. If you cannot afford the additional cost of fencing, then use inexpensive electric wire.

Horses are selective grazers and will chew down certain areas, creating "roughs" and "lawns." These then become the newest growth places and contain the tastiest and tenderest grasses. They will generally defecate in one area of the pasture and then not eat from that area, which is actually a form of self-protection from over-infestation by worms.

Knowing this can also help you plan your health-care program. Patrol your pasture area weekly, if not daily, and pick out the manure piles just as you would in a stall. It's a lot of work, but it will pay off in better health for your horses. You can have them fully graze a portion of pasture and then move them on to the next small area while the first recovers properly. The result is fewer parasites, more evenly grazed fields and better overall management.

2. Fertilize and reseed. After removing the livestock, you need to put your management skills to work as you lime, fertilize and mow the area for maximum regrowth. The best time to fertilize is in the fall, not early spring, as fall feeding directs the new energy into the roots for stronger plants. For spring feeding, go with a higher phosphorus/potassium, lower-nitrogen fertilizer to encourage further root development instead of a sudden rush of green that stresses the plant. Although an extra-green spring pasture looks nice, it will not be so healthy when droughts and grazing tax its endurance.

All soils will become depleted of the nutrients necessary for plant growth unless you resupply them. After all, the soil's nutrients go into the grass and then into your animals. To see exactly what your pasture needs, take a soil sample, have the extension office send it off to a land-grant university, and get an exact reading of your soil. In the absence of a soil test, apply 60 pounds each of nitrogen, K_2O and P_2O_5 plus 2 tons of lime per acre.

If you have an established pasture that is not being used for hay and has high levels of K_2O and P_2O_5, refertilize with those nutrients every 3 to 4 years to maintain their levels in the soil. If clover represents at least 30 percent of the plant population in the pasture, no nitrogen fertilizer is needed. But if it is absent, apply 50 to 60 pounds of nitrogen in early spring and again in late summer each year.

If your established stand of grass thins out, reseeding is called for. Do this in February or March, using a no-till drill to seed into the existing thin sod. Do not allow horses to graze these pastures until the old plants reach a height of 5 to 7 inches. Then when the old plants are grazed down to 2 inches, remove the animals again to let the new growth establish itself.

3. Control weeds. Another important aspect of pasture management is to control and eliminate weeds. You can do this by either mowing or applying chemicals, with a combination of the two being the most effective practice. Mowing must be done before weeds go to seed, so they are not spared by the mowing operation.

Chemical applications should be timed to eliminate weeds as they emerge with full leaves in the spring, but not so late that they are strong and well-established. When shopping for chemicals to control weeds, always read the labels for toxicity levels. You can poison your horses if you don't watch out! If you need in-depth information, check with the local Cooperative Extension Service for local recommendations. Commercial chemical and fertilizer companies can also provide information and custom application of their products.

A word of caution about chemicals: While there are many products approved by the Food and Drug Administration (FDA) with directions that sanction immediate grazing, it is always best to allow a heavy rain to wash the chemicals into the soil and off the grass. Most tests for toxicity apply to oral ingestion, but do not address possible damage to a horse's lungs from inhaling chemicals.

4. Remove manure. Removing manure is your best method of parasite control, but it is also the one you are least likely to keep up with over the long run. Horses will pick an area of their pasture in which to defecate and then won't graze that area. This is how pasture grasses become uneven and this condition will only get worse if nothing is done to correct the situation. Traditional recommendations used to be that you clip and drag the whole area to level the grasses and spread the piles. Current research indicates that the life span of the worm eggs is such that it allows them to lie dormant, so they can reinfest your horse, even if you allow the pasture to weather for a month or more. Some researchers feel that dragging only spreads the eggs in a wider area, thereby increasing reinfestation. Realistically, most operations do not have the option of leaving pastures unused for long periods of time.

While the most effective solution is work-intensive, it is the best for your horse's health. You should treat your pasture like a stall and pick the piles out of it as often as possible. Experiments with heavy-duty lawn vacuums and lawn brooms have not yet produced the perfect result, but a gentle stroll around the pasture with shovel and wheelbarrow can do wonders.

PREPARING YOUR SITE

No matter the shape of your land, plan to put your building on higher ground, even if you have to build it up with fill or clear some trees to get to a high spot. The goal is to create a slope away from the barn or run-in shed that will carry waste and rainwater away from its footing where they would rot out the wood as well as make your barn a sloppy mess.

It's just as important that the run-in shed drain as efficiently as does the fancy barn. Both simple and complex facilities have two common functions as far as drainage is concerned: to provide a dry place for your horse in times of wet weather and to encourage the runoff of urine. Since horses are territorial, they will want to claim the run-in shed just as they will a stall. Especially if your soil drains poorly, it's best to place the facility in the highest location.

You should be using treated lumber for your barn footings to avoid rot, but, even so, the daily wear of

horses in and around the stable can take a toll on the structure and the surrounding area. If your finances allow, it's a good practice to lay a rough-surfaced asphalt or concrete apron around the entire barn at least 4 feet from the base of the building and make it slope gently away from the barn. Too steep a slope, however, can invite accidents, so allow just enough of an angle to draw water downward.

A simple way to create a concrete apron effect is to slope a layer of stone dust out from the barn, then sprinkle powdered cement dust over the stone. Wet it liberally from the hose, or wait for a rain, and you'll find the stone dust dries as a solid.

If that is more than you planned for, or more than you want to undertake, at least run downspout drainage well away from the barn. Even barns need gutters. If pastured horses have access to the barn, you will need to horse-proof the drainage, as horses like to loaf around the corners of a barn and can crush lightweight drainage pipes where they are on or close to the surface. As with any part of the barn, just make sure your plans do not compromise your horse's safety.

One solution is to buy clay drainage pipe, 4 or 6-inches in diameter, and place a section at the bottom of the downspout, which runs into the end of a small trench leading away from the barn. Then run black corrugated plastic drain pipe, either perforated or solid, PVC pipe or more clay pipe to the end of the trench. If you are on a hill and the end of the pipe is where horses can get to it, make sure you use clay pipe where the pipe resurfaces, or the end will be flattened by the horses and the pipe will back up to the base of the barn.

If your barn is midway down a hill, level the ground above the structure or even put a shallow ditch between the drainage and your building. Then steer the runoff well away from the barn so it won't be undercut in the first big rain.

If you're building in marshy territory, you'll have to dig a series of ditches and dry wells to overcome a damp barn area. You can have these dug by an earth-moving company at some expense, or you can easily rent a small ditch-digging machine that will dig a ditch with a minimum investment of time and money. Then you can place black perforated drain pipe in the ditch which can then be covered over, thereby eliminating an open ditch in the field. If you calculate the slope of the ditch properly, the underground drain pipe will pull excess water from the area.

To put a dry well in a troublesome low spot, dig a hole at least the size and circumference of a 55 gallon drum and then fill it with brick and stone debris. Dig a ditch nearby, run a small slanted trench from the dry well to the ditch, and place a length of corrugated black plastic tubing from the top of the well to the bottom of the ditch. Use the type with scattered holes throughout. This will speed drainage in all directions.

Finally, plan your buildings with an eye to esthetics. Keep the front of the property pretty, landscaping it with nontoxic plantings. Locate the trampled areas behind the barn, and your farm will look many times more professional.

4 FULL FARM

PLANNING

It takes more than a lone barn to make a farm, and you can ease the work required to run your facility by good placement of the important extras: paddocks, rings, roads and storage buildings.

Each item that you wish to build on your property will have unique requirements as to footing, drainage and easy access by such things as delivery trucks, so first you need to decide what you are going to have, and then where you will place them.

Take a diagram of your property, either the actual site plan or something that includes the major features of the terrain, and add the fixtures to scale. This will give you a fairly accurate idea of the relationships between these features and their proportions to each other and your overall property.

Also indicate on your map the type of soils located on the property. Your SCS representative can help you map your property as to soil type and use, and may even be able to make recommendations as to the best building sites and road and pasture locations. Since soils have different profiles and thus different best uses you may realize that your first choice for a road site may be a wet, soft piece of soil at the driest of times, far better suited for pasture than for a road. And if you plan a bathroom in the barn, your soil must have even better percolation than if you planned only a drainfield.

Your goal is to get the most for your money. If you have a natural area for a road bed, why pay extra money to place the road in a location that may require tons of gravel? Only you know your overall plan, and it may well be that the lay of the land and other considerations force you to place the road, ring or barn in a certain location.

This is where your overall planning during a think-through stage comes in. Being aware of extenuating factors before the fact will keep later problems from stopping your whole operation.

Once you've located the best spots for the barn and any outbuildings, you can plan the driveways and fencing. Here your key thoughts should be access and security: access to every part of the farm for vehicles large and small and security for your horses from each other and careless passersby. Unless you are working on the tightest of budgets, or with a highly unusual site, plan from the start to enclose the barn area in its own fence, both from regular pastureland and from the highway.

Fencing the barn away from pastured horses will save you from such problems as chewed siding, rubbed corners, crushed drainpipes, and much more. Unless you plan a true loafing shed, don't let your barn be treated like one. Loafing horses turn into pests, and not just the chewing type. They cannot resist harassing another horse who, once stabled, cannot escape the annoyance of a wandering companion. Horses loose in the barnyard can also find their way into tack and feed rooms, damage parked tractors and trailers, and steal any loose grooming implement. Worse, they can escape from the barn area and onto roads and others' property. While fencing your barn will mean more expense and putting in at least one more gate than you would need for pastures alone, the security is worth it.

Before planning the gate site, draw up your preferred driveway plan. Don't think of casual visitors when setting up the driveway, but of delivery trucks. Trucks need the greatest turning room and most secure footing. Since your deliveries will be to hay and grain storage, bedding areas and perhaps manure pickup spots, the driveway should be in as straight a line as possible from the street to the end of the barn. Then add a large enough pad or backing lane so that even an 18-wheeler can make a simple three-point turn. Such large trucks may seem far removed from your plans now, but as soon as you order a few tons of hay, you'll see the advantage.

Don't skimp on the width of the driveway. Trucks and trailers track wider than a car. If the driveway's shoulder is steep, you risk a serious accident even at slow speeds if a tire should drop over the edge.

If your barn is on a slope, take the expense of grading your road into account when you plan both your barn and the outbuildings. A few dozen more feet of road can cost a great deal. Perhaps moving the barn a touch closer could save dollars better spent on a new saddle.

Managing the Neighbors

While zoning may force you to build in a certain area, no matter where you build you must consider that horses are often looked upon in the same legal sense as swimming pools and other things that appeal to young people: attractive nuisances. The very presence of a horse is like a magnet to many children, and you should keep this is mind, not only when selecting an area in which to build, but in planning the layout of your facility. Setting your fence lines several feet inside your property's borders allows you to ride around the perimeter and also buffers you somewhat from the neighbors.

Since any normal fence can be easily scaled by a child, consider placing "No Trespassing" signs in strategic locations on the property's borders. Develop a rapport with neighbors who have children so that they understand they are more than welcome to visit when you are home, but that your property and animals are off-limits at other times.

Become a good neighbor by realizing that not everyone enjoys the smell of horses or the sight of a neat manure pile. Some suggestions include:

1. Keep the overall appearance of your property up to neighborhood standards. Keep the pastures clipped and clean and the barn painted or stained.
2. Avoid trashy-looking areas, piles of used lumber or jumps, even if it is valuable material. If you must store building supplies, then do so neatly and in an obscured place.
3. Isolate manure in a confined area, and remove it regularly to hold down insect populations.
4. Have an ongoing fly and insect control program so that neither you nor your neighbors are bothered by the pests.
5. Keep dust levels to an acceptable minimum by following good pasture management and applying an economical surface treatment to your road and work areas. Again, you as well as your neighbors will benefit from this.
6. Take time to drag rings and riding trails to keep unsightly manure piles to a minimum.
7. Reduce a potential eyesore by efficiently storing equipment and machinery in your facilities.
8. Plan for areas of high traffic to have solid footing so they are able to withstand vehicular and animal movement and not create mudholes.
9. Arrange large service deliveries, such as those by feed trucks and hay tractor trailers, to be made so they won't disturb neighbors by blocking the road and making excessive noise.
10. Limit your farm pets, and provide a space to restrain your dogs. Since many dogs like to chase ridden horses, confining them is a safety factor in your best interest. Not only that, but horses seem to do poorly, in both training and health, when constantly harassed by dogs.
11. Consider putting in landscaping that is not only esthetically pleasing but also easy to maintain. Unless you like to cut grass or your covenants are restricted, keep your yard small and turn all available land into pasture.

All in all, the way you lay out your farm should make your operation more efficient and easier to work around, as well as more pleasant for you and the neighbors.

This end view of one of the barns at Paper Chase Farm in Middleburg, Va., shows how sloping land can best be used by a drive-in manure pit. A small platform extends out over the area behind the tractor, allowing manure to be dropped directly into the manure spreader parked there. The rear portion of the area provides wheelbarrow storage, as shown, so that loose or passing horses cannot injure themselves on a wheelbarrow left in the aisle. With a pit as close to the barn as this, it is wise to keep a pole across the platform area so that a loose horse cannot fall in. Photo by M. F. Harcourt.

MANURE DISPOSAL

You have no idea how much manure a big, healthy horse can produce until it snows so deep you cannot get the wheelbarrow out of the barn to dump what you have cleaned out of your horse's stall. Clearly, one of the foremost problems you have to deal with is manure storage and its ultimate disposal.

If you or your neighbors like organic gardening, then you have the biggest headache basically solved. Horse manure makes fine garden mulch once it is well aged, and with proper liming, it can enrich the poorest of soils. However, it does need to age several weeks before planting because of the high ammonia content.

Storing manure until it is disposed of is another story. Criteria for adequate manure storage should include:

1. Manure storage should be close enough to the barn to be convenient, yet far enough from the barn to avoid trouble with flies and far enough from the house not to create odors in hot weather.

2. It should be easily accessible from the barn, no matter the weather, by whomever/whatever will remove it. For example, if you are on a hill, it's easier to push a full wheelbarrow downhill and an empty one up.

3. It should be contained in some manner so that it does not become like the eggplant that ate Chicago and engulf the stable yard.

Spreading fresh manure back on pasture land, even if you plan on leaving the land vacant for several months, is not a good idea, as worm eggs of various species have great longevity and can reinfest your horse months after being spread. New studies have proven that the old theory of spreading manure so as to expose eggs and larvae to sunlight and air does not kill adequate numbers of eggs. Thus, you are only reinfesting your horses with these methods.

You have other options. First, manure can be composted in an enclosed structure (built with cinder blocks or other material not eaten away by the manure) and sold or given away in the spring to avid gardeners. For the best compost, spread a coffee can of slaked lime over each wheelbarrow-load as you go along. To keep your manure pile well-groomed may seem an odd thing, but if you enclose it on three sides and then shovel the front part well back as it grows, more manure will be composted quicker and the pile will stay within the confines of its walls.

Avoid letting the approach to the pile become fouled with spilled bedding, for instance. You don't want to create an untidy and unsafe footing in a highly traveled section of your farm. You can spread dirty bedding on pasture not intended for horses.

Most parasites are species specific, so cattle can benefit from the enriched grass you provide without picking up the worm problems your horses would get.

You can also spread manure on ground intended for crops, but, here again, check with your Cooperative Extension Service as some crops do not tolerate this well. In some areas, there are manure container and removal companies that will handle it like any other garbage—a fine arrangement, although potentially expensive.

And a last note on manure pickup: If cleaning stalls into the spreader suits your taste, and you can't wait until the horses are all turned out, it's a good practice to park the spreader attached to the tractor outside the barn and clean the stalls into a large, wheeled plastic garbage can. While these cans won't negotiate mud well, they're fine on dry ground and in the aisle. You can clean several stalls into one can, and then roll it swiftly to the spreader, or store it temporarily with the cover on.

BEDDING STORAGE

Choosing bedding of the type and amount you require depends on several factors:

1. Will your horses be fully stalled, in/out depending on weather conditions, or in only at odd times?
2. Do you clean all the stalls or do you have reliable help or hired labor to do it for you?
3. Types of bedding should be:
   - Easily available.
   - Economical and fit your budget.
   - Esthetically and otherwise satisfying.

Once again, your source of information should be the Cooperative Extension Service and local horse owners as they will know the types of bedding most available in and best-suited for your area. With today's markets, however, you need not be locked into one type. If your pocketbook allows and your storage space is adequate, you can opt for any type of bedding you want.

In addition to straw and wood chips, consider chopped newsprint sold in bags, or sand, peanut hulls and pine needles. Each has its own pros and cons, but all share one characteristic: They must be kept dry to be effective. And as a time saver, if bedding materials

are stored along the path to the manure pile, you can pick them up in the barrow on the way back to each stall.

Bedding materials that are stored in piles, such as wood shavings, require frequent truck deliveries, so you need to provide room for a dump truck's raised bed, which is ordinarily blocked by most shed roofs. To solve the problem, consider a roof that can be lifted off in sections, or one that can be raised from one side with a pulley. That way you'll avoid dragging snow-laden tarps back and forth, or worse, filling a stall with wet bedding.

Shavings on the farm of Penny and Lloyd Burger of Chesapeake, Va. are kept dry yet accessible with this roomy container. Made of 4″ × 4″ posts with plywood sheets as walls, it is roofed with a series of lift-off panels that allow stand-up or drive-in access as far back as you wish. There's no need to drag a heavy tarpaulin or move an entire cover all at once. Roofing panels can be fiberglass or corrugated aluminum over 2″ × 2″ framing, strengthened as needed as the dimensions of your shavings pit require. Photo by M. F. Harcourt.

This two-sided shavings pit features a simply framed metal cover that is hinged on one side. Pulleys are attached to the free side and to the barn near the ridge pole, allowing one person to easily lift the shavings-pit roof for deliveries. The two-sided design with raised cover, unlike a three-sided one, lets a dump truck drive through without backing to unload. (Windchase, Hillsborough, Va.) Photo by N. W. Ambrosiano

An entire shed for shavings is not out of the question if your operation is large enough. This shed combines closed storage, at right, with two-bay shavings and/or vehicle storage under a 14-foot roof. (Paper Chase Farm, Middleburg, Va.) Photo by M. F. Harcourt

INDOOR ARENAS

No matter the climate, an indoor arena is a great boon to a barnowner who has either a boarding or a training operation. Shelter from sun, rain and snow allows consistent exercising and a secluded atmosphere for working young or excitable horses. Building your own arena is a task that won't be tackled here, as the clear span—at least 66′ × 132′ for the smallest of dressage arenas—is beyond the skills of most amateur builders. But you can have either an arena company or any builder with the expertise for such large buildings do everything from just the roof and its supports to the plushest of turnkey arrangements, wet bar in the lounge, and all.

Having a steel-building contractor raise your roof has several benefits. First, you get a column-free interior, an absolute prerequisite for a safe indoor riding area. Second, the roof does not require the support of the walls, so you can either go without, or make them as weather-tight as you wish. Just remember that you are building a year-round facility, and if your weather ranges from subzero to 90°F, you must allow for both insulation and good air flow.

If you wish to finish the building yourself, your options are almost unlimited. In warmer climates, you may choose to merely fence the covered area. This gives you a working area that is sheltered from the sun and rain and still allows summer breezes to blow through it.

If inclement weather is a problem, then walls are called for, but it is possible for them to be weather-tight without creating a claustrophobic feeling. When using any exterior ply product available or noisier sheet metal over a simple 2″ × 4″ frame, be sure to add several 5′ × 8′ windows on each side and at either end. You can put a sliding door over each, or a shutter arrangement, but, either way, you have the shelter of the indoors with the light and air of the outside. A panel of light fiberglass-reinforced plastic or plexiglass at the top of the walls will give you a much lighter arena as well, without compromising your weatherproofing.

If you plan to insulate walls and ceiling, note that swallows, pigeons and sparrows are among the many squatters who will noisily move in if you don't prevent them. Stretch wire mesh across every bit of insulation or you'll have a bird-condo from wall to wall with such noise that you won't be able to hear a trainer from 10 feet.

No Footing, No Horse

The footing indoors is as subject to wear and tear as that in your outdoor arenas, even if it doesn't get direct rain and wind. Many types of special arena footings are sold through advertisements in horse magazines, or you may have your own favorite. Most of them combine wood shavings with another material, such as sand or peat, for springiness, freeze-proofing and reduced dust. Companies that supply footing materials can dump the basic material for you to install, or you can have them prepare the footing from below ground level on up.

The essential ingredient in all these footings is a good base that drains well and neither sinks deep nor packs down to a stonelike surface. There are as many combinations of sand and gravel suggested by experts as there are experts, so preparing your footing gives you a prime chance to talk with local excavators, agricultural advisors and other arena owners about what works best in your region.

No matter what kind of top dressing you put over the base, it, like your horses, will take some grooming to be at its best. Lack of rain produces a dusty arena that leads to choking, so you'll need some irrigation plans. If you don't mind spending several hours with a hose in your hand, you can try hand-watering. A step above that is installing lawn sprinklers upside down along the rafters, which provide a more automatic arrangement (for example, you can set these on timers to sprinkle an hour before the first ride of the day). This also gives you some wet patches, though, which are unacceptable in a potentially perfect riding surface.

The smoothest watering devices have copper tubing with tiny holes every foot or so that are laid out in a grid or rows along the rafters the length of the arena. They are operated, like any sprinkler system, from a main valve and work best if you have good water pressure to force an even mist over the full area.

Dampness alone won't give you great footing though. It can be overdone, and in cold weather it can make even the most freeze-proof surface pretty icy. Under such conditions, limit your watering to moisten the top few inches of the surface; whatever you do, don't drench it heavily with water.

Raking or dragging is the other big job required to

maintain arenas, especially those with heavy use. Hand-raking all the way around the edge is effective, but it uses time you'd probably rather spend riding, pulling manes or even cleaning tack. If you have a farm tractor, or even a small car with a trailer hitch, you can help maintain any footing by running a harrow or shallow cultivator over it every week. This breaks up clods, erases trenches and keeps the footing aerated well below the surface, which is essential to protect your horse's legs.

Circular, rotating harrows with short teeth are wonderful for arenas, but they require a three-point hitch on a tractor, as do most liftable harrows. Simple drags, while not as sophisticated, can do a decent job and don't require the tractor hitch. A chain harrow, for example, has a mesh of heavy links with protruding fingers on one side and a smooth surface on the other. Held square by a heavy pole at the front, it then has a chain that hooks on to any pulling vehicle so it can be dragged rough or smooth-side down, depending on the final polish you want on the arena's surface.

In a pinch, a sheet of wire-mesh or chain-link fencing will work, provided that it's weighted down with a railroad tie or two and held square at the front. It does not provide deep conditioning, but only smooths out the rough bits.

Extras

With basic walls, footing and conditioning planned, you can consider the nonessentials. For a start, a sloping interior wall, 4 feet high, will keep your horse from fracturing your kneecaps as it attempts to evade your inside leg in the course of your schooling sessions. A gentle slope moving outward from bottom to top will ensure that as the horse's feet near the bottom wall, you will still have a foot or so of space before your stirrup begins dragging along the higher wall. This interior wall must be made of heavy boards or thick plywood, as a horse misbehaving or turned loose can kick it in passing and might force a leg through it with disastrous results.

Along other safety lines, don't allow any objects to protrude into the arena, such as jump poles stored alongside, door latches that stick out, or hooks to hang your training equipment on. The walls at horse and rider height should be completely smooth and safe. And don't forget to provide a tractor-wide door to the inside, not just a horse-width one.

Lights are essential for any arena with walls, or even for open-sided arenas where people like to ride after dark. You have to invest in some powerful hardware, as they must be hung fairly high in the air, not require constant maintenance, and spread light evenly across the floor. A walled-in arena gets very dark very quickly, even if you install clear plastic panels along the eaves. Don't hang the lights lower than 15 feet from the floor, and if you plan any jumping indoors, take them up to 20 feet or more to be safe.

Check your local suppliers for their best buys on lights big enough to spread their illumination over a wide area, not spotlights. And remember that you're lighting a dark-walled, nonreflective room, so increase your estimated wattage requirements to account for that. Placing sodium-vapor street-type lights every 20 feet or so is effective, but check with local builders for their recommendations.

It's good to have a viewing gallery, no matter what your purpose is in building the arena. It can be as simple as an 8' × 16' judges' stand that raises judges 3 feet off the ground, or as elaborate as a heated seating area complete with popcorn and plush seats. With a viewing area, buyers can get a better view of "the product" and students can see better examples. If the viewing area is closed off but has a window, discussions and comments won't intrude on the rider's concentration. If you have a window, though, make sure there's a part you can open to call instructions to your riders.

BREEDING SHEDS

A special breeding area is essential to some farms, whether a fully booked season calls for all-weather servicing or mere modesty requires an enclosed area. Like a riding arena, a breeding shed needs good, dust-free footing, plenty of headroom and good lighting. More than that, though, it needs a wash stall and teasing area to prepare both mare and stallion. The wash stall can also double as a lab and examination station for the vet.

Solid construction is essential for any teasing barriers and breeding chutes as well as plenty of padding on exposed corners and edges. Prefabricated metal chutes may be ideal for you, or perhaps you may want to design your own.

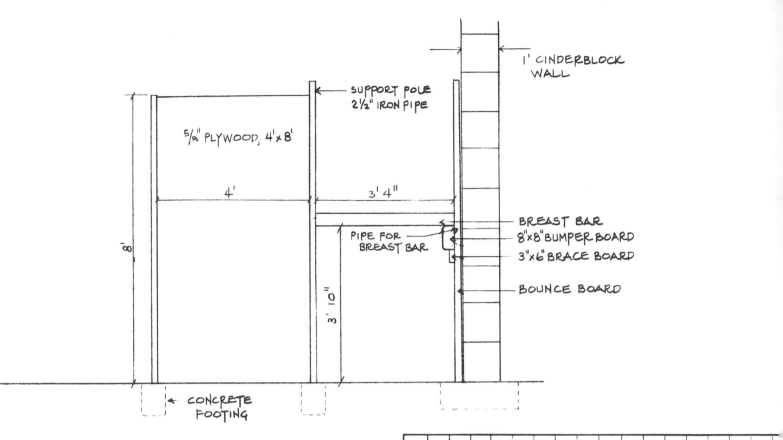

SUPPORT POLE
2½" IRON PIPE

1' CINDERBLOCK WALL

⅝" PLYWOOD, 4'x 8'

4'

3' 4"

8'

BREAST BAR
8"x 8" BUMPER BOARD
3"x 6" BRACE BOARD

PIPE FOR BREAST BAR

BOUNCE BOARD

3' 10"

CONCRETE FOOTING

Seen in cross section, the breeding chute is of simple but elegant construction, protecting both horses and handlers.

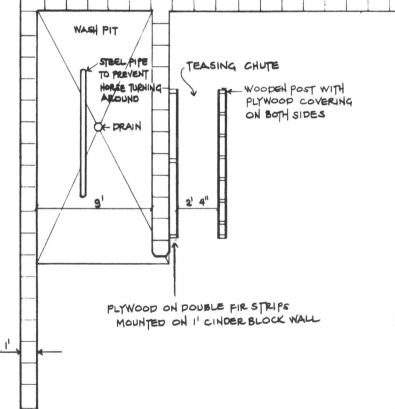

WASH PIT

TEASING CHUTE

STEEL PIPE TO PREVENT HORSE TURNING AROUND

WOODEN POST WITH PLYWOOD COVERING ON BOTH SIDES

DRAIN

9'

2' 4"

PLYWOOD ON DOUBLE FIR STRIPS MOUNTED ON 1' CINDER BLOCK WALL

1'

BEVELED EDGES

The breeding area, below, is designed for safe, two-person operation. The mare handler stands protected behind the short, angled wall of the breeding chute (at far right in sketch), holding the mare's head as she faces the breast bar. The cinder block exterior wall is padded with a shelf running the length of the chute, supporting the stallion's right side as the stallion handler stands to his left. The padded rail also prevents crushing the stallion's foreleg against the solid wall. At left, a wash stall to prepare mares and stallions for breeding is located within the main breeding shed. Immediately adjacent to both the teasing and breeding areas, it saves time and effort for a small staff. (Courtesy Hassell Arabians, Reddick, FL)

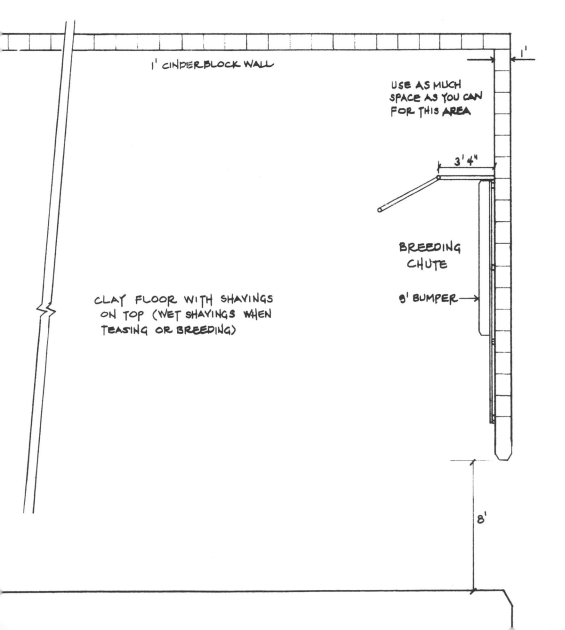

1' CINDERBLOCK WALL

1'

USE AS MUCH SPACE AS YOU CAN FOR THIS AREA

3' 4"

BREEDING CHUTE

8' BUMPER

CLAY FLOOR WITH SHAVINGS ON TOP (WET SHAVINGS WHEN TEASING OR BREEDING)

8'

This is a good example of a workable trailer, hay and run-in shed. (Captain and Mrs. D.A. Williams, Chesapeake, VA.) Photo by M.F. Harcourt. See page 152 for complete plans and specifications for this structure.

TRAILER/GENERAL STORAGE

The variety of vehicles and support equipment that goes along with horses is surprising. Storage of this equipment can be simple but at the same time, if properly organized, it can add years to the life of your machinery. In storing your trucks, trailers, tractors and so on, you not only can protect them but can ensure safety by not keeping them where your animals can injure themselves.

All vehicles that use combustible fuels should be stored in facilities not adjacent to your barns. First, there is the increased danger from fire due to the fueled vehicles, and, second, there is the damage that exhaust can cause to your horses' respiratory systems. Even using tractors for manure removal while the horses are in the barn is not advised. If it is absolutely necessary to do so, then all air passages should allow maximum circulation while work is underway.

In building an equipment shed, you'll find that the sizes that worked for horses are almost perfect for vehicle widths, but not lengths. An average tractor is 6 to 8 feet wide, with a length of 12 to 16 feet. A manure spreader can be as narrow as 5 feet, but rarely wider than 7, although its length can be as much as 16 feet. A truck needs 8 feet of width and 24 feet of length, and a trailer needs at least that. Be sure to add extra space needed to open vehicle doors, load and unload saddlery and conduct minor maintenance.

Any kind of equipment, such as a manure spreader, should be stored so that the horses cannot come in contact with parts that may cause injury. This goes for lawn mowers, small wagons or any other heavy equipment. Implements used in your daily cleaning operation in the barn itself should be stored/hung up out of the traffic patterns of both horses and humans for additional safety. Be sure to store your wheelbarrow in an out-of-the-way location.

A trailer shed to hold these things, just a simple pole building with optional sides, can also double as hay storage either in the loft or on the floor area. Here, the hay's dust particles will do no harm, and you'll get two types of storage under one roof.

Fencing

The possibilities for fencing are almost endless these days, since the technology of plastic, vinyl-and-wire, rubber and metal fencing has been developed well, and there's always good old wood.

Good fences truly do make good neighbors, or they at least keep you from having one more reason—loose horses—to disagree with the neighbors. You need an enclosure that is secure, economical, durable, and will not hurt errant livestock. Those requirements used to mean one thing—post-and-rail fence or a combination of post and wire mesh. While that is still a highly popular method of fencing, new materials are gaining popularity because of their durability, ease of installation and safety.

Depending on your horses, you can pick from the fences within your price range—slip rail or board, post and rail, wire mesh with a board or electric wire along the top, rubber, smooth high-tensile wire, wire with vinyl "rails" attached, metal poles, PVC poles or rails, or plain electric wire. They all work.

The only wire you should never have anywhere near your horses is barbed wire. They can and will kill themselves on such a fence. Even the quietest old campaigner's instinct is to run until stopped and then struggle when tangled. Combine those instincts with a fence that breaks the skin at the slightest touch, and you have mangled horses.

Here is a general list of fencing costs and materials for a square 5-acre paddock (1,868 sq. ft.) Board and woven-wire fences have posts spaced at 8-foot intervals, while high-tensile wire posts are located every 30 feet, with droppers at 10-foot intervals. No labor costs are included, and material costs vary with terrain, source of material, and so on.

Fencing Materials and Costs*

Fence type	Materials	Costs
4-board	232 line posts	$1,058
	1 end post	11
	464 oak boards, 1″ × 1″ × 6″ × 16′	1,856
	50-lb nails	58
	1 12-ft pipe gate	55
		$3,038
		$1.63/ft
		$3–$4.50/ft installed

Woven wire, board on top	16 4″ × 8′ braces	65
	228 line posts	1,040
	5 end posts	55
	16 9″ dowel pins	8
	8 4″ dowel pins	2
	16 twitch sticks	16
	116 oak boards	464
	12-lb nails	14
	50-lb staples	50
	16 100-ft rolls of woven wire	1,895
	1 12-ft pipe gate	55
		$3,664
		$1.96/ft
		$3.10–$4.50/ft installed
High-tensile wire 8 strands	16 top braces	65
	50 line posts	203
	5 end posts	49
	8 5″ × 8′ brace posts	58
	8 4″ × 8′ brace posts	32
	16 9″ dowel pins	8
	8 4″ dowel pins	2
	16 twitch sticks	16
	104 grooved droppers	161
	7 in-line stretchers	21
	1 tension spring	5
	1 pkg wire sleeves	13
	20 lbs staples	20
	4 4,000-ft rolls wire	269
	1 12-ft pipe gate	55
		$976
		$0.52/ft
		$1.60–$1.95/ft installed

Note: The 1988 Federal Tax Law allows for fencing as a capital investment if your operation qualifies as a business, and thus your fence can be depreciated.

This table was compiled by James Dunford, farm management extension agent, Culpeper, Virginia.

This combination fencing provides high-tensile wire set in channels along a PVC plastic "rail" that flexes for safety, as horses bounce off the smooth, highly visible fence. It even comes in colors that make it resemble classic wood fencing. Once installed correctly, this fencing has an almost indefinite lifetime, according to its manufacturers. As with standard high-tensile fencing, you can set the posts much farther apart than the standard 8- to 12-foot increment, thus saving on post costs. (Kafri Farm, Orange, Va. and Breaking Dawn Stables Quarter Horses, Keswick, Va.)

PVC post-and-rail or post-and-pole fencing is slightly more expensive than wood rail fencing, but has a far longer lifespan. Sun-blocking chemicals in the plastic have overcome the weathering problems this type of fence had several years ago, and it now is accepted as a fine fencing option. Its flexibility makes it a good fence for use with rambunctious livestock, and it never needs painting. (Triple Crown Fencing)

A strong, creosoted 4-rail fence keeps foals from slipping beneath the bottom board. Its height and solidity discourage horses from jumping or pushing it. Photo by M. F. Harcourt.

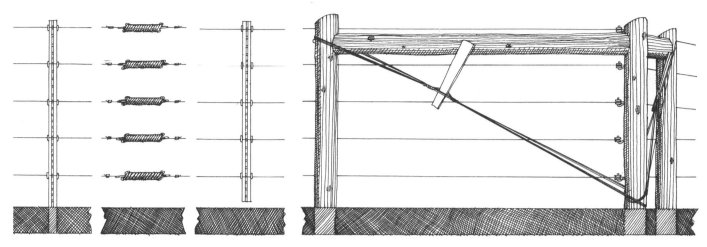

A high-tensile wire fence is excellent for large properties. The long distance between support posts makes it economical, and, once the corner braces are properly set, the fence lasts a long time. Any or all of the wires can be electrified to keep horses off and intruders out. (Techfence, Advanced Farms Systems, Bradford, Me.)

Rubber fencing has received little press since its initial splash into the market during the 1970s, but it makes a wonderfully economical, long-lasting fence. At installation, some important steps are required in addition to firmly placed posts. A propane torch is needed to seal off edges where loose strings may attract chewing horses, and a facer strip of fencing run down each post helps hold things together longer. (CFM Inc., Marysville, Ohio)

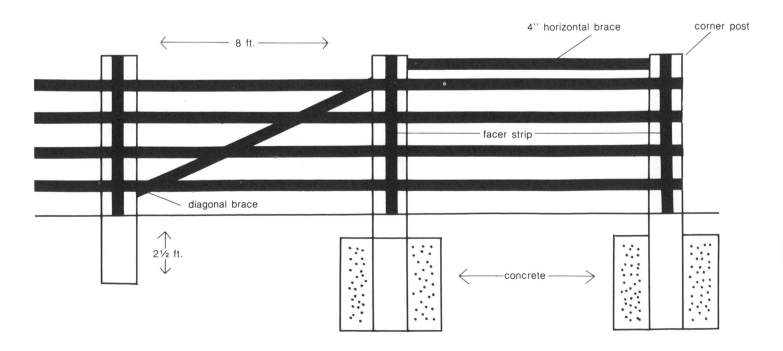

Split-rail fencing is handsome and secure in areas where it is available. Left unpainted or stained, it eases labor. Like post-and-board fencing, it is excellent for areas of rolling terrain or uneven pasture borders. Photo by M. F. Harcourt.

For regular use, a 3-board post-and-rail fence, nailed with the boards on the horse's side, is excellent. Like 4-board fencing, this can be used for winding fencelines and on hilly terrain as each panel stands independently. (Tokaro, Middleburg, Va.) Photo by M. F. Harcourt.

Diamond wire fencing, topped with a rail or a strip of electric fence, is a superbly safe type of fence. Its diagonal reinforcements are impervious to the smallest hoof, making it safe for breeding farms, and, like other wire mesh products, it can be found in heights up to 6 feet. Photo by M. F. Harcourt.

Stone walls, in regions where stones abound, are handsome but not the safest for horses. They shift over time, becoming easy to climb over, and a careless horse playing in a field can be injured by crashing into such a wall. If your farm has walls, simply line them with an electric wire to keep horses clear, or run a separate line of fencing inside the perimeter. Photo by M. F. Harcourt.

You can add to the lifetime of your fence by following some simple steps. First, set every post as deeply and firmly as you can. This means tamping the soil to a concrete-hard texture around every single post. If your soil is damp and rots wood fast, soak the bottoms of all your posts in creosote for at least a week before setting them in the ground, and then surround them with a sandy mixture to let the area drain well. Even better, substitute PVC or metal posts for wood in marshy areas and take advantage of their impervious properties.

If you're using a fence that takes tension, such as high-tensile or rubber, follow the manufacturer's directions to the letter, and your fence will stay up and functional far longer than if you hadn't. With these fences you sometimes need an extra tool or two, such as a stretcher or special tension-holders, but they are worth the expense. A fence set correctly the first time is always stronger than one you have to fix up along the way.

With all but post-and-rail sections, you should plan to set at least the corner posts, or sometimes every post, in concrete. That is not hard to do, but it does add to overall fencing costs. Other incidental costs to consider are shipping fees from fence manufacturers to you, specialized fasteners for synthetic fencing, and paints, stains or coatings for wooden fences. When you install your fence, place the posts on the outside of the fencing materials so horses cannot run into the hard edges of the post.

A hint on keeping board fencing at its best: If using oak, don't paint it. Oak lasts twice as long if left to weather gently on its own. Other woods benefit from a good coat of paint or stain applied when the fence is dry. To keep rot away, leave the bottom edge of each board bare. That allows trapped moisture to escape.

Apply paint with a brush or sprayer, coating each board thoroughly. Missed spots let in rain and the board will rot inside the paint coat. When using creosote, which has limitations on who can buy and apply it, you can get the fastest coverage with the type of heavy-duty, 4-inch-wide roller used in applying textured paints. These hold up best on the rough fence surface. Be sure to wear rubber gloves, goggles, and a face mask when applying any of these chemicals, as none of them is good for you. Some will react with the sun on your skin as you work, giving you a nasty rash, too.

5 WHAT TO PLAN FOR

INDOORS

LIGHTING

One of the things that can make a barn either a pleasure or a plague to work in is the lighting. For those interested in anything more than the most cursory of grooming and management chores, a well-lit view of the scene is essential, and not just during the brightness of the midday sun.

Natural lighting not only makes the facility more comfortable and enjoyable for you but also for your animals. They will be more cooperative when handled if the surroundings are open and have adequate lighting. Horses have an intense natural fear of close, dark places. That is why it takes time to train them to go into trailers and they are hesitant about new areas they cannot see well.

Skylights or windows in the roof can add large amounts of natural light, and choosing models that open has the added benefit of providing extra ventilation on demand. Skylights can range from the simplest of corrugated, fiberglass-reinforced plastic sheets, to small domes, to complex window packages.

If you choose to install clear panels, place them directly over key working areas, such as the main aisle in which you plan to groom, or over each stall, as the light will not be diffused very far from its source. If you have them installed by a professional, insist on some guarantee that they will not leak, as any perforation in the roof is likely to do so unless perfectly installed. Plenty of flashing and/or caulking is essential for a tight seal around each nail or screw and around the edges.

If piercing the roof is not to your liking, add a clear plexiglass border at the top of the walls, just below the roof, to take advantage of low sun rays in the winter. This skypanel will be amply shaded in the hot summer when the sun's angle is more directly overhead. Adding this clear border all the way around dramatically increases your natural lighting, so long as you do not extend your roof's overhang too far out.

Even if you decide not to install lights inside the stalls, you'll need electricity for lighting, running radios, clippers, water heaters and vacuums. Having lights during the dark, cold winter months will allow you extra working time after an early sunset—time you'll need for blanketing, bandaging tired legs and chipping ice from buckets.

The type of lights you install is fairly flexible as long as you place them for maximum effect. Both incandescent and fluorescent lights are safe, and the

A transparent "skypanel" at the top of the walls allows additional light to enter the barn. If you plan 8-foot walls with plywood siding, you will need to cut the top 2 feet off all your siding, a potential waste if your budget is tight. To avoid this, use board and batten, shiplap or some other siding, or plan 10-foot walls to give your horses safer headroom. This picture also shows an excellent sliding window cover, one that can be opened or closed from inside the stall and that won't blow in the wind. (Joe Ann Scott's barn built by Upperville Barns, in Aldie, Va.) Photo by M. F. Harcourt.

An illustration of how one might place fluorescent lights between stalls so each stall gets light from two directions. By hanging the lights over the partitions, there is no need to buy two lights per stall.

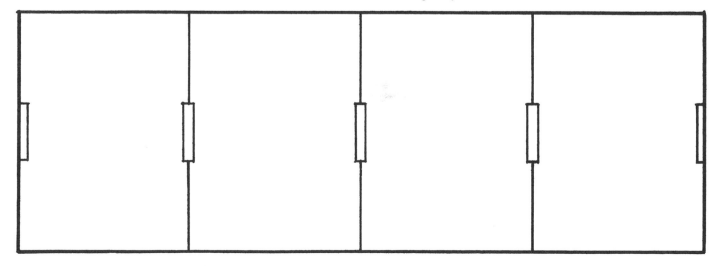

horses don't care. If you choose to install one light per stall, place it, well-shielded, in the center of the ceiling, not in one corner or along a wall. A horse casts a large shadow, effectively blocking a large part of a stall when there is a side-mounted light.

For even better working conditions after sundown, install two lights, placed opposite one another along the walls, so in the clear light of morning you won't find your horses halfway clipped or groomed. If you run your stall dividers up partway but not all the way to the ceiling, you can place fluorescent shop lights between each stall and get the benefit of two lights per stall, for nearly half the cost of two lights in each closed stall. For example, in a four-stall barn, you'd have to install eight lights to have dual light sources in each. If you instead use partial dividers, five lights can be split between the stalls and still give you light from two sides in each.

Having one area extremely well lit for routine close work or emergency vet work is most handy, as most emergencies seem to occur after dark. If you need to keep a horse well groomed or prepare one for competition or other activities, then having a well-lit area in which to groom/braid or prepare for anything will be useful. If you need to set up a foaling stall or wash horses during cold weather, you might want to use heat lamps, so plan your wiring so it can be suitable for carrying the required amount of current. Both 110 and 220 come in handy in at least one outlet in the work area.

While lights may seem to be the only electrical need in the barn, consider the number of things you may wish to plug in from time to time, such as horse clippers and vacuums. An outlet of the exterior type, with a spring-loaded cover nearby or in each stall can be useful, as are outlets around the tack room and general work areas. Other potential uses for electricity in your facility include: bathroom heaters, hot water tanks, washers and dryers, intercoms, video cameras, security systems. In fact, just about anything you can imagine could be added to your facility,

depending on what you want and how much you want to spend. The key is to put in as much electrical capability as your budget allows—and then some.

All in all, your minimum needs include:

- One light every 10 feet of aisleway
- One double electrical outlet at each end of the barn
- One light in each stall
- One well-lit area for grooming/vet work

As for the outside, a floodlight over one or more barn doors is a lifesaver if you feed after dark or unload after a show. You can install a simple floodlight with an easily located switch, or you might look into getting a light-activated light, such as a street lamp, which will brighten the area as soon as the outdoor light drops below a certain level.

When installing electrical fixtures, do it right the first time. Unless you really know what you're doing, you'll have more peace of mind if you know that the wiring and electrical outlets have been safely and correctly installed by a professional, thereby reducing your chances of a barn fire. Be sure to enclose all lines, both interior and exterior, in metal conduit, protecting the wires from the predations of mice, horses and sharp objects. Horses are hopelessly curious about wires crossing their walls, especially foals and horses on stall rest, and one chewed wire can kill the horse and burn down the barn.

Install a switchbox that allows for ample future expansion. Adding sheds, stalls and work areas will be less of a nuisance if lighting is prepared ahead of time. Inside the barn, recess all light fixtures, if possible. If not, be sure to cover them with a sturdy grate or screening to protect them from breakage and possible electrocution of either people or animals. Place them as high as possible; you may not think fixtures are low until one is broken by a rearing horse or a rake or shovel being used in its vicinity.

VENTILATION

Adequate ventilation in your facility means providing a clean, fresh exchange of air in your barn. By that we mean fresh air, not drafts. Loose construction, with no means of shifting or shutting off the breeze, can aggravate illness to the point of pneumonia. Good, planned air flow is necessary not only for

your horse's health but also for yours and that of any employees you may have.

Natural ventilation can be used in most cases by laying out the structure to take advantage of the prevailing winds, correctly spacing doors and windows, and adding any cupolas or vents you may feel are

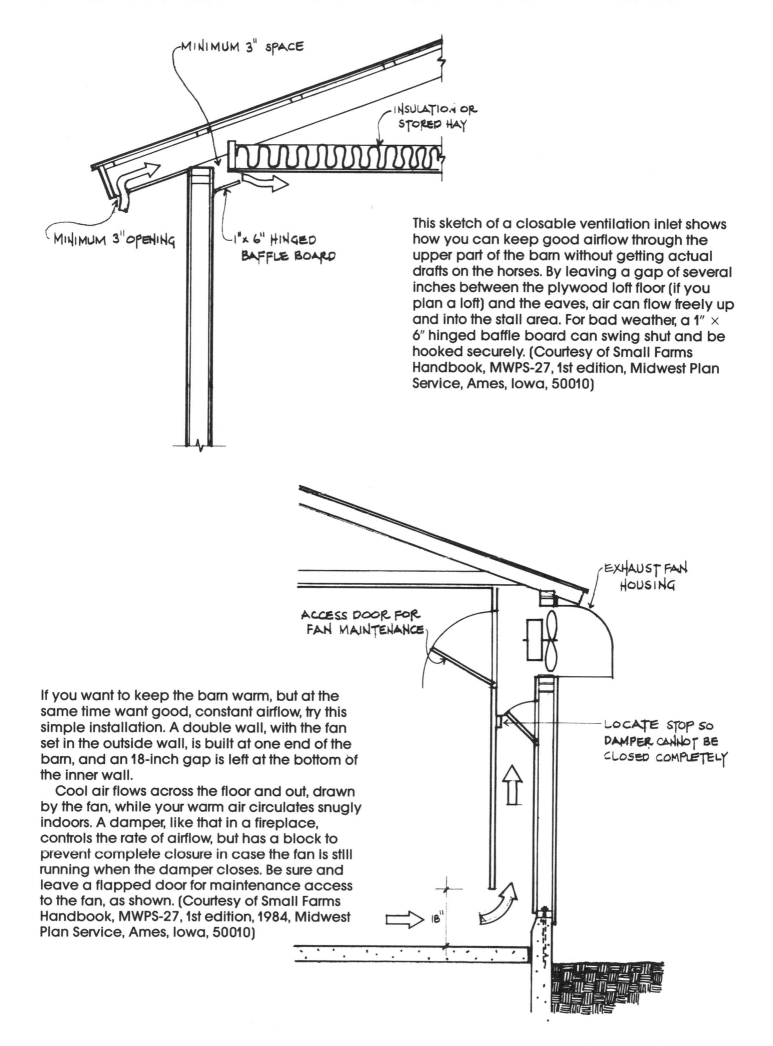

MINIMUM 3" SPACE

INSULATION OR STORED HAY

MINIMUM 3" OPENING

1" x 6" HINGED BAFFLE BOARD

This sketch of a closable ventilation inlet shows how you can keep good airflow through the upper part of the barn without getting actual drafts on the horses. By leaving a gap of several inches between the plywood loft floor (if you plan a loft) and the eaves, air can flow freely up and into the stall area. For bad weather, a 1" × 6" hinged baffle board can swing shut and be hooked securely. (Courtesy of Small Farms Handbook, MWPS-27, 1st edition, Midwest Plan Service, Ames, Iowa, 50010)

EXHAUST FAN HOUSING

ACCESS DOOR FOR FAN MAINTENANCE

LOCATE STOP SO DAMPER CANNOT BE CLOSED COMPLETELY

If you want to keep the barn warm, but at the same time want good, constant airflow, try this simple installation. A double wall, with the fan set in the outside wall, is built at one end of the barn, and an 18-inch gap is left at the bottom of the inner wall.

Cool air flows across the floor and out, drawn by the fan, while your warm air circulates snugly indoors. A damper, like that in a fireplace, controls the rate of airflow, but has a block to prevent complete closure in case the fan is still running when the damper closes. Be sure and leave a flapped door for maintenance access to the fan, as shown. (Courtesy of Small Farms Handbook, MWPS-27, 1st edition, 1984, Midwest Plan Service, Ames, Iowa, 50010)

18"

necessary for your climate. Here's where a knowledgeable local or the county extension office can give you tips on how the extremes of climate and prevailing wind patterns affect horses in your area. They can also offer suggestions for aligning your facility most sensibly on your property by taking the climate into account.

If you can manage good air flow without extra fans, so much the better. Some steps can go a long way, such as adding 6 to 8-inch spacing along the eaves, using a cupola or inserting ridge vents the entire length of the building. You want the hot air to rise, pulling in cool air on the floor level and spreading a cool breeze along the way that's essential in hot, muggy weather.

In particularly close, hot climates, vents along the stall wall can help. Placed about 2 feet above the floor, a long, narrow slot vent, 2″ by 7′, can keep the stall floor much healthier, especially for foals who don't benefit from air flow that's only at window level. Adding skylights that can be raised also allows heat to escape.

If you need to put many horses in a small area, be aware that each animal puts out a considerable amount of body heat and airborne contaminants. Research has shown that horses' lungs are noticeably damaged by the effects of dust, mold, ammonia and other pollutants in the air. Owners of performance horses and breeding stock should be especially aware of this, as a horse's performance suffers greatly from diminished lung capacity. Increased sickness in the barn can also be attributed to bad ventilation.

To clear the air with more than passive measures, you can install fan and venting systems large or small. You may only need to place attic fans in the eaves above the stalls, or you could tackle the problem the way the professional agricultural folks do. They have access to good information and plans for adding mechanical ventilation systems, often at surprisingly reasonable prices. Here again, your County Extension Agent can offer suggestions, many of which are simple enough for almost anyone to incorporate into existing building plans.

WATER

Two main factors will affect your plumbing plans in the barn—your water supply and what you hook up to it. It makes everything you do in and around your facility easier if your water source is both adequate in supply and efficiently laid out.

The purpose of having a water supply is to provide water in adequate amounts to maintain your horse's good health. Horses require a large quantity of water —5 to 15 gallons of it—on a daily basis. Major health problems result from a shortage of it over just a few hours. Climatic conditions as well as exercise and carrying a foal can increase the normal amount a horse consumes, so providing water a horse can drink whenever he's thirsty, day and night, is a necessity.

Hot weather increases a horse's consumption of water, just as in people, but few realize that cold weather also puts an increased demand on your horse's water supply. Winter feeds are usually drier than summer grasses, and a horse needs even greater amounts of water to compensate for dry feedstuffs.

The size of your water supply depends on the size of your operation, the number of animals you care for, and the type of operation you have. The simpler your operation (a run-in shed for a pleasure horse, for example), the simpler your water needs. Of course, if you add the requirements of horse bathing and blanket washing, the amounts increase proportionally.

First, decide if you need to dig a second well or if your existing well will support both your household and your barn. If you are on a city/county water line, check to see about the increase in cost if you use large volumes of water. This may have bearing on whether you install a barn-only well or merely practice great efficiency in stable chores that require water. Finally, if you drill a well for the barn, be sure to have the water tested for chemical and bacterial contamination.

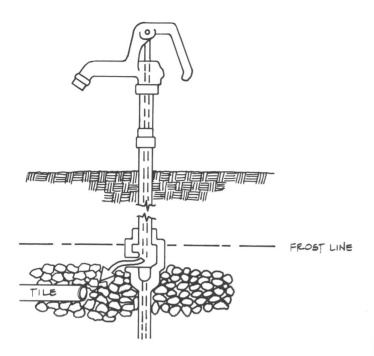

This sketch of a frost-free hydrant shows the placement of the drain-back valve below the frost line. Be sure to prepare the underground drainage area well with gravel and tile to allow runoff away from the barn. (Courtesy of Small Farms Handbook, MWPS-27, 1st edition, 1984, Midwest Plan Service, Ames, Iowa, 50010)

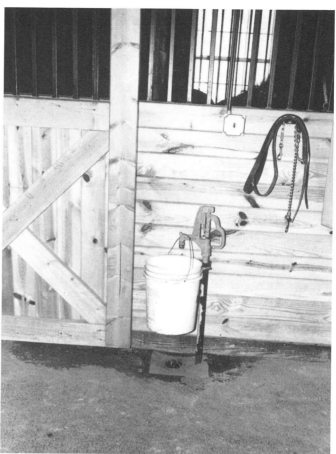

A water hydrant in the aisle has an underground drain designed to catch overflow. Be sure the drain covers you choose with a system like this are removable for cleaning clogs, but can also be firmly attached to keep hooves from slipping into the hole. Photo by N. W. Ambrosiano

Plumbing Installations

When planning plumbing installations, one of the decisions you will have to make is whether you want automatic waterers. They are convenient and can save you time and effort in keeping an adequate supply of water in front of your horse. They come with or without heaters, and the better models are splash- and spill-proof, effectively defeating the efforts of bored horses to flood their stalls.

Like buckets, automatic waterers need cleaning on a regular basis. And they are not without drawbacks: Most importantly, you cannot keep track of the amount of water your horse consumes on a daily basis. By mentally recording the daily consumption of water, you can catch colic and other digestive problems before they become serious.

Automatic waterers can also be broken, usually in the middle of the night, by an errant kick from your horse and thus flood the barn. In chilly climates, even heated ones can freeze, as the water lines to them are not turned off. Thus it is recommended that you use metal pipes to supply waterers, so that you can thaw frozen sections with a blow torch. Metal supply lines also prevent horse teeth from opening sections unexpectedly. To minimize frozen pipes in hard-to-reach areas, wrap supply lines with electric heater tape except for sections the horse can reach.

If you plan on faucets as your water source, drain-back valves are the most reliable type to use. They will give you a long period of service with little or no maintainance and will minimize those annoying mud puddles that form around water supplies. These

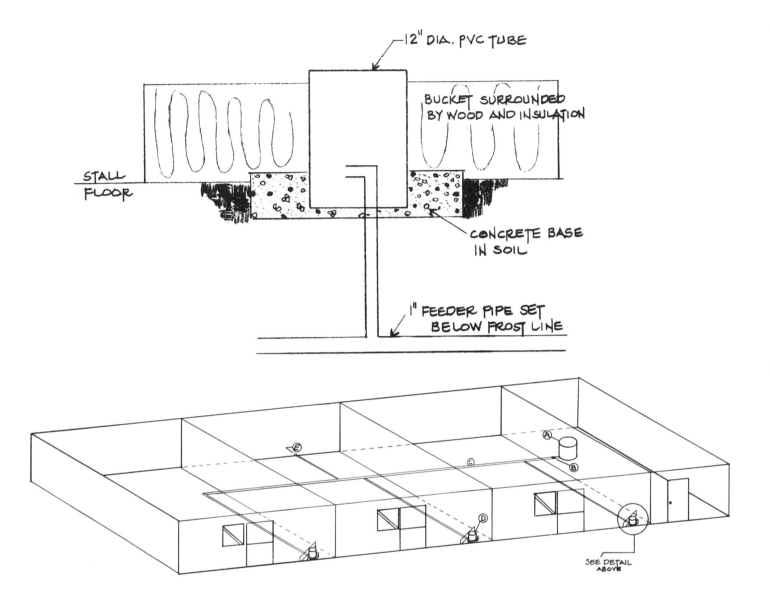

This watering system allows for gravity flow to each bucket and is frost-free as well. A heated, main holding tank in the tack room, kept filled by a simple toilet tank-type floater valve, is the heart of the system. From there, PVC pipe runs below the frost line the length of the barn, branching to each stall's corner waterer.

The "buckets" are actually short sections of 12-inch PVC pipe set into a concrete pad through which the feeder pipe runs. Each is set inside an insulated corner box and is leveled to the height of the main tank's water level. No part of the feeder pipe is ever exposed to air or horses' teeth.

To drain the system, a separate drain pipe and valve branch off the main pipe. With the closure of the main tank's valve and the drain valve's opening, all the waterers can be flushed out at once. Before flushing the system, one should be sure to pick out extra debris from the buckets so that it doesn't block the pipes on the way out. A quick swish around the inside with a scrub brush cleans off the sides. If you have a wet/dry vacuum, you can suck the remnants out of each waterer in moments. The model shown was designed by Roy Rottenberry and Grace Dawson, and is installed in the barn of Mrs. Dawson and her daughter Phyllis, an Olympic Equestrian Team member, at Windchase Farm in Hillsborough Va. While this design is "homemade," other designs can be purchased from some farm supply centers.

The gravity-flow "bucket" is shown in its insulated corner box. Notice that there are no protruding objects for a horse to break or hurt himself on. A simple interior frame holds the front of tongue-in-groove 1" × 6"s with a piece of plywood forming the surface through which the PVC pipe protrudes. The interior of the box is filled with non-toxic foam insulation. Photo by N. W. Ambrosiano

The main water tank and heating unit are major components in the gravity-fill watering system. A hinged wooden lid coated with spar varnish keeps the tank from being contaminated or a danger to farm creatures. Photo by N. W. Ambrosiano

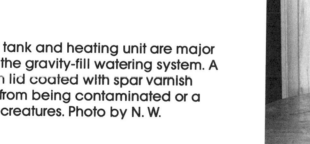

The main tank with the cover lifted shows the floater valve with shutoff coming from the heating unit, upper right. Below the surface is the open end of the main pipe leading to the buckets. At this end of the pipe, a shutoff valve allows the system to be sealed for drainage from the other end. Photo by N. W. Ambrosiano

valves, instead of letting water lie in the pipe to where the faucet is, automatically release water into a small below-ground drain field when the faucet is turned off. That way, no water sits in the pipe above-ground, where it may freeze. Drain-back valves can be installed in several locations in the barn and in the pastures to provide water under the most severe weather conditions. Just remember to check the freeze line in your area (depths range from 0 to 2 or 4 feet underground). Then install the drain valve well below the freeze line.

If you need hot and cold running water, then you will want to include a hot water tank in your plans and suitable plumbing to handle the additional line. This is not complicated, and it is one measure that will make much of your horsekeeping easier. The small tank is relatively inexpensive to buy and run and if it is a quick-recovery type it can provide plenty of hot water for a minimal investment. This will allow you to add a washer and dryer to keep all the towels, wraps, blankets and such clean and avoid wear and tear on your household appliances.

Additional water users, such as bathrooms and wash pits, are wonderful additions, but not essential. If you think you might want to install these later, arrange the plumbing so that fixtures and associated structures can be added. Just be sure that the runoff from these will not overload your current septic system. Before you decide on an additional bathroom, check your county regulations, zoning permit and percolation permit to be sure it is permitted.

Just as with electrical wiring, put in as much plumbing as your budget will stand. Even if you don't add the bathroom, washer and dryer or wash pit right away, put in the line and connections. Remember to keep diagrams of the lines, valves and fixtures for future maintenance reference.

Obviously, you don't need all this fancy stuff to enjoy your horse, but if you are into showing, eventing, hunting or training in a big way, then these things make it easier and quicker to keep your equipment and horses clean and well-maintained.

FLOORING

Before you finish your stalls and fill them with bedding, remember that they will take a great deal of abuse from the combination of a horse's feet and the inevitable waste products of so large an animal. If the floor isn't tough, you'll soon have a bog in the barn, no matter how deep the bedding.

The flooring needs to provide good drainage, easy manure removal, and minimal dust and noise. It's hard to find all that in one surface that most people can afford, but there are some options: Mats over concrete, packed earth and even straight sand may fit your needs. It's a matter of considering what you have available and how you'd like to arrange your stalls.

For a start, drainage in stalls or a shed can be helped by laying a mini-drain field in each stall or by adding a mini-drain field running through the center of your stalls. To do this, in the center of the stall dig a hole about 3 feet in diameter and deep enough to reach a layer of dirt that will drain adequately. Then fill the hole with large-diameter gravel or alternate layers of sand and gravel and stamp that firmly into place before covering it with the dirt you have selected for the shed/stall. Another good, simple way to prepare a stall is to lay a 6- to 12-inch deep layer of

fine gravel or stone dust, and then place your dirt, clay or other materials over it.

If you find a layer of bedrock below your top soil, then you can use it as the "bottom" of a mini-drain-field by channeling urine out of the barn and allowing it to drain down and away from the facility. Not that much is really going to drain, as the bedding will absorb much of the moisture overnight.

Porous "popcorn" asphalt is another flooring that works fairly well. It provides a solid covering for the drainfield over which you can spread a good layer of bedding to protect a horse's tender joints from the hardness and roughness of the surface. To get a good surface for horses, have the hot asphalt raked rather than hot-rolled during installation, and when you order it be sure to ask for the type with large particles.

Another fairly costly but well-supported flooring method is to place a grid of pressure-treated 2″ × 4″s edgewise across the stall over a 12-inch base of gravel and fill stone dust between the boards. The boards should be between 1½ and 4 inches apart, braced at either end, and the stone dust should be packed firmly between them to the top of the board edges. This drains superbly, provides better traction and more give than a concrete floor, and allows those cleaning the stall to use a shovel, which can be slid along the boards for the last part of the cleaning.

A simpler drainage option involves packing the flooring soil in a gentle slope to one corner or to the rear of the stall. Then plan to clean that damp area thoroughly and lime it weekly to prevent odors and bacterial buildup. When grading the flooring, keep the slope to a maximum of 3 (degrees) so that your horse's legs will not become strained by standing at such an angle for long.

You have another choice in the final layer of dirt that lines the stall. Classic flooring is hard-packed clay, which offers a tough, yet forgiving surface that won't scrape your horses or cause their legs to swell if they must stand on it for many hours. Thick clay is the best, packed to a nearly hoof-proof surface, because it lets urine run off to a drain hole and only needs to be reworked once a year or so. Some people order baseball diamond clay for this and pack it no less than 4 inches deep. If no clay is available, any thick, packable soil free of stones and sandy portions (which disintegrate) will do, but it may require more frequent repacking.

To form the most solid soil flooring, you need a heavy pounding tool with a flat foot and a waist-high handle that you can raise and drop repeatedly while moving slowly around the stall to level the entire surface. Motor-driven settlers for concrete do the best job, provided you have both a hose and a supply of loose clay on hand to alter the dampness or dryness of the surface as you go along.

If you plan to wash and disinfect your stalls, as in a breeding or hospital arrangement, more efficient drainage is in order. In some operations, actual drains are placed in the corners of the stalls and a solid flooring such as concrete is covered with commercially sold rubber stall mats. Most farm managers lay bedding over the mats, removing it when the stall is disinfected and replacing it with clean straw or wood chips before the horses are brought back in.

Others prefer not to place bedding over the mats at all, washing the stall contents into the corner drains daily. The drains empty into a sluicing system that can be flushed weekly and thus the stalls are kept clean and fairly odor-free. Since the only waste material is pure manure and not bedding, stall cleaning is easier and there is less of a manure pile to be disposed of. The pure manure can be sold or given away as organic compost.

The drawback to using matting in your stalls with

Use a tamping tool, as shown, to pack down your floor soil adequately. Be sure you work the floor thoroughly from one side to the other. Photo by N. W. Ambrosiano

no additional bedding is that in cold climates the facility may be colder than you like. Also, it's hard to beat the esthetics of a barn well bedded with gleaming straw or well-banked wood chips.

Stall mats can be expensive, but they are not hard to find in most areas. You can also order mats from one of the many companies advertising them in horse magazines, or find a company in your area that might have used conveyor belts on hand. The thicker grades of old belts make fine stall mats when they are cut into 10- or 12-foot lengths.

If you live in an extremely wet area and cannot afford to build your barn well above the water line (summer rains in coastal southeastern states almost demand that you keep your horse on a second floor or buy one with webbed feet), you can depart from the firm flooring concept completely. In some operations, native sand is used because it drains rapidly as stall/shed footing. The corresponding management technique is never to use bedding, letting the sand be the bedding and removing droppings daily. As the sand becomes cleaned out with the droppings or toted out on the horses' hooves, more sand is simply added. In a moderate to warm climate, it works well. In a climate where cold temperatures demand dry bedding, this would not adapt as well.

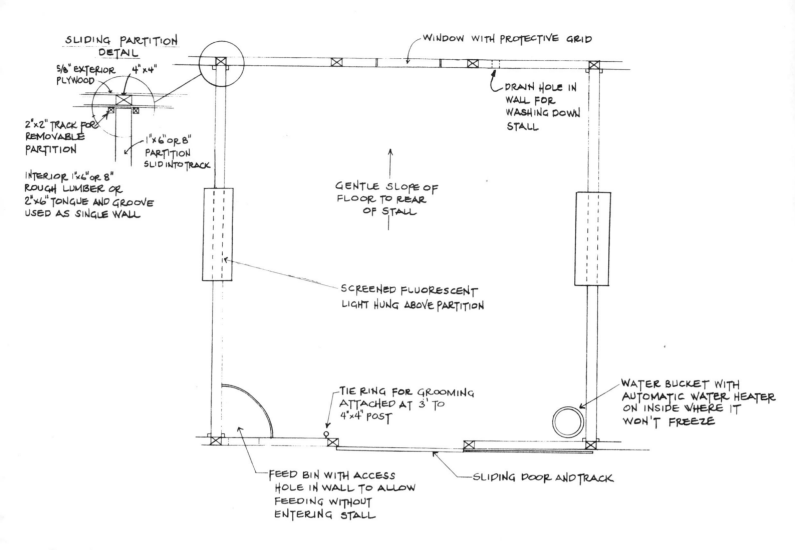

SLIDING PARTITION DETAIL

5/8" EXTERIOR PLYWOOD   4"x4"

2"x2" TRACK FOR REMOVABLE PARTITION

1"x6" OR 8" PARTITION SLID INTO TRACK

INTERIOR 1"x6" OR 8" ROUGH LUMBER OR 2"x6" TONGUE AND GROOVE USED AS SINGLE WALL

WINDOW WITH PROTECTIVE GRID

DRAIN HOLE IN WALL FOR WASHING DOWN STALL

GENTLE SLOPE OF FLOOR TO REAR OF STALL

SCREENED FLUORESCENT LIGHT HUNG ABOVE PARTITION

TIE RING FOR GROOMING ATTACHED AT 3' TO 4"x4" POST

WATER BUCKET WITH AUTOMATIC WATER HEATER ON INSIDE WHERE IT WON'T FREEZE

FEED BIN WITH ACCESS HOLE IN WALL TO ALLOW FEEDING WITHOUT ENTERING STALL

SLIDING DOOR AND TRACK

Sample stall layout, showing good placement of feed and water buckets, tie rings and lighting.

STALL DESIGN

Stall Size

Based on the size of horse or pony you are housing, you have some options in stall size. The smallest comfortable stall for an average horse is 10′ × 10′ which allows just enough room for the animal to turn around in and lie down safely. A 12-foot area is more generous for the animal and gives you more working space if you choose to do your grooming in the stall. However, consider that in enlarging your stalls you increase your materials—bedding, roof area, everything—by 44 percent or more. It's a simple calculation of area: a 10′ × 10′ stall is 100 square feet. A 12′ × 12′ stall is 144 square feet, needing almost half again as much bedding, roofing, and so on.

Owners of stallions or broodmares really have no choice, as breeding operations must allow space for mares to foal and stallions to live comfortably. They need stalls that are 12′ × 12′, 14′ × 14′ or larger.

A simple way to construct a foaling stall on short notice is to remove a partition between two regular stalls. Removable partitions can be made by sliding 1″ × 6″ or 2″ × 6″ boards down between a channel of boards (2″ × 2″s or 4″ × 4″s are fine) on opposite walls. Convertible stalls are also useful when you have a horse on extended stall rest.

Constructing all your stalls 12′ × 12′ with removable partitions between each allows your facility a wide range of uses and isn't that hard to do in the initial construction stages.

If you have planned a hayloft, (try not to, as hay stored in a shed is safer, easier to use and won't fall

The aisle of this barn is kept safe and clear because it has properly installed sliding doors: heavy-duty latches, a bottom roller to keep feet from getting between door and wall, and a stop block at the rear of the door's track allowance. You can build such a door for yourself, buying the hardware at farm-supply stores and rigging a 4′ × 8′ door panel of plain exterior plywood (as heavy as you can find) with a Z-frame of 2″ × 6″s. For a more polished effect, back the plywood with varnished tongue-in-groove hardwood (shown here), with T1-11 grooved plywood, or elongate the traditional X pattern with six 2″ × 6″s nailed in place. Insertion of a commercial stall grill as shown is optional. (Cooper's Farm, Aldie, Va.) Photo by N. W. Ambrosiano

through cracks onto your horses' coats and into their lungs), allow enough clearance so that no part of the supporting structure is lower than 10 feet. That way, your horses will not injure themselves when they throw their heads in excitement or fear. If you are working on a tight budget and will have smaller horses in the barn, 8-foot ceilings are acceptable, but be sure to completely recess your light fixtures.

Doors

Standard stall doorways should be 4 feet wide. Once the doorway is framed, however, you have choices as to what you fill it with. Light or heavy doors, screens, chains or web barriers are all appropriate, depending on your animals and the time they spend indoors. If you wish to save expenses during construction, consider using inexpensive stall guards as a start-up door. Later, you can add either preconstructed doors or doors you make yourself.

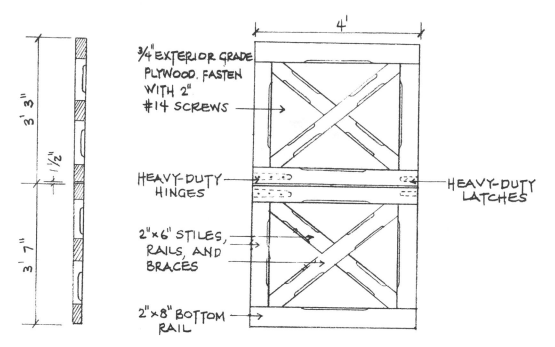

³⁄₄" EXTERIOR GRADE PLYWOOD. FASTEN WITH 2" #14 SCREWS

HEAVY-DUTY HINGES

2"×6" STILES, RAILS, AND BRACES

2"×8" BOTTOM RAIL

HEAVY-DUTY LATCHES

3' 3"

1½"

3' 7"

4'

To build a traditional Dutch door, you need simple, solid materials such as 2″ × 6″s, ⅝ths-inch plywood, heavy strap or Tee hinges and metal edging for the chewable sections. While this door plan shows a full 4-foot door, if you are building with 4″ × 4″ posts set 4 feet on center, you will need to make this door 44 inches wide, not 48, so it will fit neatly between the posts.

Sample stall specifications show a sliding door installed with treated lumber near the floors, commercial stall guards, and partitions made of planks set between 2″ × 2″ channels on the walls. (Small Farms Handbook, MWPS-27, 1st edition, 1984, Midwest Plan Service, Ames, Iowa, 50010)

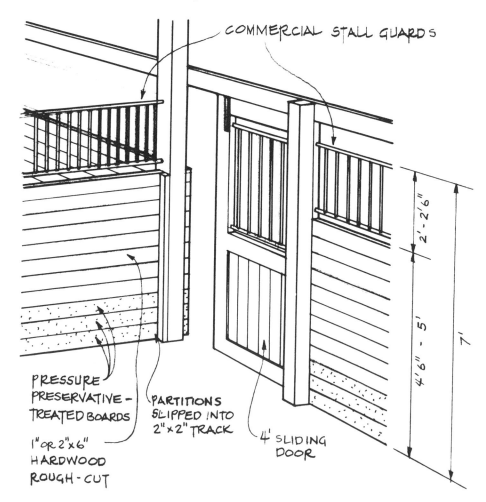

COMMERCIAL STALL GUARDS

PRESSURE PRESERVATIVE-TREATED BOARDS

1" OR 2"×6" HARDWOOD ROUGH-CUT

PARTITIONS SLIPPED INTO 2"×2" TRACK

4' SLIDING DOOR

2'-2'6"

4'6" - 5'

7'

Solid swinging doors can either be the Dutch types (separate upper and lower parts) or a full door which may be solid or have steel mesh or bars on the upper portion for light and ventilation. You can make standard Dutch doors fairly simply, provided you measure each section carefully and conduct several trial installations as you go along. The trick is to have both doors hang securely and evenly so they do not swing open or closed on their own, striking the horse unexpectedly. And they must never swing into the stall, as they can crush the horse against the frame as he passes or become stuck in the bedding as they swing.

A sliding door may be more expensive initially but it adds aisle space to your facility and is generally safer. It's much harder, for example, for a horse to get a hip hung on a sliding door than one that swings into his path as he passes.

Doors can also be purchased as complete, ready to hang units. Most are sold in standard 4-foot widths, whether they swing or slide. Search the horse magazines for the many companies advertising these and other barn materials, but be aware that shipping

Used with or instead of traditional doors, heavy stall screens are available from livestock catalogs and through some feed stores. They come in many sizes and shapes and are excellent for providing light and air while keeping horses securely in their stalls. If using full-length ones such as those shown, be sure to hang them high enough so that a horse's hoof cannot slide under and get caught. Install these with a simple screw eye in the door frame for the hinge pins to drop into. A double-end snap at the other side also attaches to a screw eye for horse-proof closure. Photo by N. W. Ambrosiano

A horse-proof and effective latch has been welded on this metal-edged sliding door. The hasp will take a lock, if necessary, while a snap attached to a chain is adequate for daily use. The rail visible along the front of the door makes a useful blanket rack, with no sharp corners on which to bump a horse's head or hip in passing. The card holder below the blanket rack allows index cards with each horse's feed or medication instructions to be changed as needed. (Windfield Station, Nicasio, Calif.) Photo by N. W. Ambrosiano

charges add substantially to the expense. In assessing whether to build or buy doors, consider purchase and shipping price versus your own time and effort combined with the cost of the materials.

Heavy gratings are a good option when hung with or instead of standard doors. They come in squares a horse can look over, tall rectangles that keep horses from biting passersby, or rectangles with a scooped opening. They are readily purchased and easily installed on eyebolts or even as sliders. Again, be sure to place them for safety so they open to the outside of the stall.

The solidity of what you install is important. Due to its frequent daily use, a door takes more abuse than the rest of the barn structure. Where lighter lumber may be adequate along a partition, use no less than 5/8-inch plywood and 2″ × 6″ boards for the door's framing and backing. A frustrated horse, either pawing or plunging against a door, can test the strength of your construction to its limits. This is especially true if you have a boarding operation or a breeding facility.

Along the tops of the doors, as on every possible chewing surface, place a thin sheet of metal to discourage destructive teeth. Throughout the stall, one hopes to have few places where a chewing or cribbing horse can get a grip, and door and partition tops are inevitable targets.

To foil equine teeth, buy drywall corner beading at a builder's supply store. It's very cheap, tooth-proof and has holes predrilled so it can be screwed to the door. Alternately, buy 4 or 6-inch wide rolls of aluminum at a hardware store, and tack it down with roofing tacks or other wide-headed nails. You need to unroll a length of metal exactly the size of the surface it will cover, cut it with tin snips, center it on the surface, and firmly bend down the metal on either side of the door or partition top. It should form a three-sided channel that is smooth and clean-looking, with no untrimmed edges to snag passersby. As it wears, remember to keep an eye out for loose tacks that can catch a passing horse or fall out only to be stepped on.

The hardware on which to hang the door is almost more important than the door itself—if the hinges sag or tear free, the door will be impossible to open or close safely and easily. Use the largest strap hinges, whether plain or ornamental, that can support a nearly 100-pound door for the best wear. The latch should be horse-proof, withstanding the efforts of nearly prehensile horse lips to open it and escape. A determined horse can easily lift a chain over a nail, flip a hook, or untie a rope. If you install the type of stall bolt with a handle that you raise and slide to free the bolt, be sure to put a screw eye through the center of the handle and attach a heavy snap to it so the horse cannot work a lip under the handle and lift it.

If your horse is prone to pawing at the door, a kick latch at the bottom as well as a top bolt will help. You can order toe-operated latches through farm or stable supply catalogs.

If your barn has a center aisle, doors to the main entrance give you the option of increased climate control. Here again, you must decide if swinging doors or sliding doors fit better with your overall facility plan. Swinging doors require more space to open and can bang or flap themselves to pieces in big winds if not securely fastened. Also, barn doors that are large enough to cover the opening of your barn often require maintenance as they age due to both their extreme weight pulling on the hingeing apparatus and the effects of the wind. If you use swinging doors, plan on heavy-duty hinges and hardware so you don't have to keep rehanging them every few years.

Sliding doors offer space-saving advantages as well as good climate control. They can be left partway open for a bit of light and air without having to be propped awkwardly ajar, and, with proper installation, they can survive high winds.

Whatever type of door you eventually decide on for anywhere in the barn, be sure that your doors:

1. Open wide enough to allow you to transport machinery, feedstuffs or other equipment through them.
2. Are wide enough for you and others to safely lead animals through them.
3. Open flush with the adjacent wall spaces so that you and the horses don't hang hips or whatever on them as you walk through.
4. Have adequate drainage near them so that the areas of high traffic do not become mud puddles during inclement weather. You may have to add some special footing in doorways to reduce mud problems. Rubber mats work well, either bought specially for stalls or made from old conveyor belting from a local concrete company, for instance. Concrete or gravel in front of the doors and up to the front of the stalls prevents the soil from wearing down to a muddy trench.

Inside the stall

Because the inside of the stall is going to take lots of abuse from your horse, plan to install just a few key fixtures. Protruding objects such as hay racks and storage boxes are an invitation to chewing and your horse may get injured bumping into them. Better to have removable stall fixtures or round the corners on those you cannot do without.

The most natural arrangement for feeding hay is to throw it on the ground. Horses are designed to eat from below their body level, and this way no dust or fragments fall down into their eyes, ears and nose. However, to keep down parasite infestation, you must keep a good clean stall if you feed off the ground. If you have an inveterate stall walker who tramps his hay into the bedding, try using a hay net (hung at the horse's eye level and removed when empty) or install a round-cornered hay rack in a corner where he's less likely to catch a hip or his head in passing. Place it so the horse doesn't get a faceful of dust and hay as he eats. Avoid the temptation to put a rack at floor level, even though it's a good height to eat from. The danger of the horse getting caught in it is too great to risk, even if you consider your horse the calmest of creatures.

Feed bins offer many choices. Good corner models are available in every price range, most can be removed for cleaning, and some with rims prevent eager horses from spilling the grain onto the floor. Buy one, rather than building one yourself: the wear of the horse eating will soon turn up splinters, the grain worked into the wood encourages wood chewing, and you cannot clean or disinfect a homemade box efficiently.

Your horse needs at least an ounce of salt per day, and a salt brick is the easiest way to satisfy him. Two types of racks are commonly available—a cage with plastic-coated wire or a metal plate with edges that catch the side channels in the brick. Both are adequate, although once the channels break off the brick from wear, the plate model is useless. You can always keep the salt brick in the feed bin, though, no matter what shape it's worn into. Or you can buy the 20 or 50-pound model and set it on a board on the floor in the corner. It will last a very long time.

For watering, you can install an automatic waterer in the corner, or simply hang a rubber or plastic bucket from a ring in the wall. Because buckets freeze in the winter, a rubber bucket is easier to pop the ice out of, although they are heavier. Rubber is also less prone to split from thermal shock, such as when you pour hot water in it on a cold day. When hanging the bucket, remember to use snaps or links that have no protruding points on which a horse can catch his face or halter.

Here's a tip the Dutch use for keeping horses watered on frozen nights: After the evening feeding and watering, twist a handful of hay into a short rope and set it in the water bucket so that one end sticks above the water level. Then, in the night as the horse goes to drink from the now-frozen bucket, he can tug the ice free as he fiddles with the hay handle.

If you plan to groom your horse in the stall, you can manage quite well by putting the lead rope over your horse's withers and letting him move as you groom. If you find yourself spending more time saying, "stand still" than grooming, you might place a tie ring in the center of one wall at the horse's eye level. Be sure to tie your rope to a loop of easily broken hay string, though, because a panicked horse in a space as small as a stall is at his most dangerous—you can't get out of his way.

FEED STORAGE

Grain storage should be close enough to your main stall area to be convenient but separated from your horses' reach by a barrier of some kind. Horses are famous for opening feed rooms and storage cans and eating enough to kill themselves.

The storage area doesn't have to be fancy, just secure. Grain can be stored in everything from garbage cans to built-in bulk storage containers lined with metal to keep out rats and mice. Grain storage should serve these purposes:

1. Easy access for feeding convenience.
2. Protection behind a horse-proof barrier.
3. No access to rodents and insects so they can't spoil grain.
4. Preservation of freshness, free from mold or mildew. These two, in addition to rodent and insect residue, can cause illness and even death in horses.

One of the most efficient grain storage containers is the plain, old metal garbage can, provided that you run a heavy spring or chain through the side handles and over the top to make it horse- and rodent-proof. A 30-gallon can holds 100 pounds of most grains and can be cleaned, moved and even painted as changes in stable management demand. To lengthen the life of these cans, set them on wooden pallets or a sheet of plywood unless they are on a concrete floor, as moisture beneath them will eventually rust a hole in the bottom.

To provide a more polished look to your feed room, you can build a counter high enough to set the cans under it, and slide each one out as you need it.

Or run two boards along the floor beneath the counter on which to set the cans. Make the back board higher than the front one to allow the can to angle slightly outward for easy grain scooping. If you like, you can then set a third board against the front legs of the counter just high enough to keep the cans from tipping over, but low enough that empty cans can be lifted out.

If you can lay hands on an old chest freezer, that makes a very secure grain bin, but you must dispose of the original door lock if there is any chance at all a child might be in your barn. Bins full of grain are magnets for children, one of whom might become trapped with tragic result. Replace the door lock with a metal hasp that is installed upside down—with the flap on the bottom—so it must be lifted into place to lock.

Some extra features in the feed room can be worthwhile. The counter, mentioned above, gives you a working surface for mixing grain and supplements, pounding medicine boluses and general work. A light and an electrical outlet are great pluses, too. The outlet allows you to plug in an immersion heater to make hot water for mixing bran mashes or scrubbing buckets.

(Right) In this simple feed and equipment area, metal garbage cans hold grain ready for feeding while bags not yet stored away are safe from dampness on a concrete floor. Tool hooks on the walls store cleaning equipment safely out of the way and give the barn a workmanlike appearance. (Tokaro, Middleburg, Va.) Photo by N. W. Ambrosiano

(Above left) If bulk feeding is your plan, a chute such as this could be ideal. Placed in the loft, the main feed box, made of galvanized iron, pours grain into a cart when the hatch is opened. The cart is then wheeled down the row of stalls and grain is scooped from there, saving the labor of carrying many buckets up and down the barn. Another advantage with this arrangement is that there is never a large supply of grain within a loose horse's reach. (Windchase, Hillsborough, Va.) Photo by N. W. Ambrosiano

(Above right) This simple 6' × 12' feed area takes half the space of a potential stall, leaving the same sized space next door for other storage. The counter here allows for mixing supplements (shown on the shelf above the counter) and general work; plus it leaves plenty of space below for storage. Note the feed scale hung in easy reach, allowing precise measurement of any feed additives or medicines. (Joe Ann Scott, Aldie, Va.) Photo by N. W. Ambrosiano

HAY STORAGE

Hay storage is a vital part of farm planning, whether you have 1 horse or 100. Even if you choose to buy five bales at a time from the local feed store, you need a good place to keep them while they last. Your storage area need not be rodent-proof as does a feed room, but it must offer some protection from water—both rain water and soggy ground. To prevent mold problems from underneath the bales, find some wooden pallets or build a simple grid, strong enough to support your weight, to use as flooring beneath the bales. The decent ventilation provided by this will keep your hay fresher longer.

Remember that hay is a fire hazard as well, so line your storage area with fire-retardant materials that will hold back flames for at least one hour. You can buy ⅝-inch fire-scored sheetrock from building supply stores; check with them on other locally available materials.

In planning the hay area, take into account the amount of hay you will buy at a time, both now and as far in the future as you can forecast. Figure on volumes of approximately 40 bales per ton, which take up about 260 cubic feet of storage space.

Buying hay directly from a farm can be most economical if you know how to select quality hay and what a realistic price is for your area. Here again, your local Cooperative Extension Service can offer education on the subject and a list of suggested hay growers and sellers.

Hay, bedding and feed storage space requirements.

HAY, BEDDING, AND FEED STORAGE SPACE REQUIREMENTS

Material	Weight Per Cubic Foot in Pounds	Cubic Feet Per Ton
Hay—Loose in shallow mows	4.0	512
Hay—Loose in deep mows	4.5	444
Hay—Baled loose	6	333
Hay—Baled tight	12	167
Hay—Chopped long cut	8	250
Hay—Chopped short cut	12	167
Straw—Loose	2-3	1000-667
Straw—Baled	4-6	500-333
Silage—Corn	35	57
Silage—Grass	40	50
Barley—48# 1 bu.	39	51
Corn, ear—70# 1 bu.	28	72
Corn, shelled—56# 1 bu.	45	44
Corn, cracked or corn meal—50# 1 bu.	40	50
Corn-and-cob meal—45# 1 bu.	36	56
Oats—32# 1 bu.	26	77
Oats, ground—22# 1 bu.	18	111
Oats, middlings—48# 1 bu.	39	51
Rye—56# 1 bu.	45	44
Wheat—60# 1 bu.	48	42
Soybeans—62# 1 bu.	50	40
Any small grain*	Use 4/5 of wt. of 1 bu.	
Most concentrates	45	44

(Courtesy: American Plywood Association.)

*To determine space required for any small grain use wheat (60# = 1 bu.) for example.
Then: 60 (4/5) = 48# wheat per cubic foot volume. To find number cubic feet wheat per ton, Then:

$$\frac{2000\# \text{ (Wt. of one ton)}}{48\# \text{ wheat per cubic foot volume}} = 42 \text{ cu. ft.}$$

Here is a properly installed hay rack, with no sharp points to hurt a horse. The built-in tray on this model allows legumes such as alfalfa to be fed without the tiny leaves falling through into the bedding below. You can tell from the height of the stall grating on the left that this hay rack is at a horse's approximate eye height, low enough that dust and particles will not fall into the horse's face as he eats. (Tokaro, Middleburg, Va.) Photo by N. W. Ambrosiano

Sources of hay from least expensive to most expensive include:

1. Hay purchased directly from the grower.
2. Hay purchased from a middleman who buys in bulk from farmers in a wide area and sells in bulk locally.
3. Hay purchased from a local feed store.

While feed stores always charge a delivery fee, sometimes both growers and middlemen will also deliver for a small additional fee. If you pay someone else to unload and store the hay it will add to the cost. Remember that growers and middlemen prefer to deliver in large amounts such as 1 to 10 tons.

If you do your homework, you can find a reliable, cost-effective source of hay that fits your overall operational plan. Check with other horse owners to see if a number of you can buy several tons at one time. Then a tractor-trailer load (10 to 15 tons) can be delivered to a central location with each person responsible for the immediate transport of his or her own share to their farm.

Remember, too, that hay is a seasonal commodity. It's cheaper in the summer when haying weather is good than in the dead of winter, so plan your storage accordingly. If you use a local source, make sure he or she has enough to sell you throughout the winter or will reserve hay for you to purchase during the winter months.

Let's add a word about management here. Through education, your Cooperative Extension Service can help you decide the type and amount of hay you need. Not every horse needs the most expensive hay. Knowledge on the subject will keep you from spending money unnecessarily and help you maintain your horse in good healthy condition.

These trap doors in the hayloft are centered over the hay racks in the stalls below, allowing hay to be shoved easily through to two horses in one effort. The doors are drawn open with heavy string that runs from the door through a ring in the rafter and tied at the end with a large washer, so you never need scratch around the loft floor looking for the door handle. In addition, the paths to each trap door in this barn are marked on the loft floor in red paint, so as new loads of hay are brought in they can be stacked out of the pathways. (Windchase, Hillsborough, Va.) Photo by N. W. Ambrosiano

As mentioned earlier, hay should not be stored above your animals' stalls if at all possible. Overhead storage is a fire and health hazard that can be eliminated by simply having a separate storage area specifically for hay (and bedding or other flammable necessities). Studies have shown that dust particles play a large role in the upper-respiratory infections often seen in today's stalled animals. No matter how high the quality of the hay you purchase, there will be some fine dust and particles that will add to the other airborne particles present in your facility. Also on a practical note, anyone running a barn soon notices that getting hay 10 feet up into a loft is no fun, even if you have a hay conveyor available.

Nevertheless, if you are planning to go on with overhead storage, then here are some precautions to take:

1. Fit the boards or plywood sheets in the hayloft floor very tightly together to reduce the amount of dust and hay particles that rain down on your animals.
2. Do not make the loft floor and stall ceiling two separate structures in the hope that this will eliminate the fall of residue. All you are doing is creating the ideal spot for a fire to start as dust and debris collect between the floor and ceiling.

While feeding from overhead racks may be convenient, studies have shown that hay stored in racks above the horse's normal head height may allow fine particles to be inhaled or to settle in the horse's nasal passages and lungs as he eats. This, like hay dust filtering through the ceiling, leads to a higher incidence of upper-respiratory problems.

TACK STORAGE

Your options for equipment storage are almost unlimited as you plan your barn. If you choose to devote all your space to horses and not horse gear, you can build all stalls and reserve a spot in the aisle for a storage trunk. On the other hand, you might choose to have a tack room that doubles as a trophy display area and hide the dirty gear in a special cleaning room reserved for scrubbing after every ride.

If everything, such as brushes, tack, and so on, has a place, then it's easier to keep everything in its place. With a more efficient operation, you spend more time riding than you do looking for hoof picks, for instance.

For the average horse owner, one room will suffice, with wall-mounted saddle and bridle racks for clean leather and a corner devoted to soiled saddlery. A tack room the size of a stall is fine if you can plan it; it will hold as many saddles as you can put racks on the walls. If you'll be the only one working in it, you'll have room for a work table and more.

Key ingredients for a workable tack room are off-the-floor storage for your saddle and bridle plus a spot for brushes and another for first aid equipment. Then, if the rising trend in tack thievery bothers you, make sure you can lock your tack area securely, and then remember to do so. The rest is gravy. There are many more wonderful things you can add, when you're designing your own facility:

- Simple hot and cold water taps, a double-sided sink, a washer and dryer for tack and saddle-pad cleaning, and an electrical outlet which allows for a coffee maker, and even a microwave for quick meals on the run.
- A saddle rack that flips to hold a saddle upside down so you can really clean the underside.
- A wide counter, where you can take a bridle apart and lay it out for cleaning, oiling and repair.
- A heavy-duty sewing machine so you can renovate worn blankets on the spot.
- A bandage-rolling spool for winding up freshly washed leg wraps.
- Two bridle-cleaning hooks, one for dirty bridles and one for freshly oiled ones to dry on.

- A nonflammable heating system for winter days.
- An air conditioner (no more moist, moldy tack).
- A refrigerator for keeping medicines, ice packs and the odd soda chilled.

Any horseperson can give you a list as long as your arm of good ideas, things that would make good horsemanship not just a matter of conscience and safety, but of comfort as well. When you tour people's barns, ask to see their tack areas, and find out what might work for you, too.

This tack-cleaning area has it all: a washer and dryer for wet pads and wraps, cupboards for storage of cleaning gear, and plenty of hooks and racks for the saddlery. (Tokaro, Middleburg, Va.) Photo by N. W. Ambrosiano

This is the side opposite the tack-cleaning area, showing both clean tack with dust covers in place and a just-used saddle, center, that awaits cleaning. (Tokaro, Middleburg, Va.) Photo by N. W. Ambrosiano

For an almost self-sufficient tack arrangement, this heavy-duty sewing machine is the final touch. Set up with plenty of work space, it means worn blankets and gear can be repaired right away. (Windchase, Hillsborough, Va.) Photo by N. W. Ambrosiano

In a boarding barn, an open, group tack room is not an efficient option. It's better to go with a row of tack lockers such as this; a floor-to-ceiling space with double doors that allow full access by the owner. Fitted with wall and door storage racks, the space is more than adequate for one horse's basic equipment. (Windchase, Hillsborough, Va.) Photo by N. W. Ambrosiano

Here is another example of a row of tack lockers.

While aisleways should be clear and horse-proof, some clever storage allows access to essentials at all times. In this barn, a horse in the grooming/wash stall need not be left alone, as everything is nearby. Exercise wraps and cottons are in the open racks on the walls, saddles wait on the folding racks at center, blankets are folded on racks along each sliding door, and the hanging mesh basket near the horse's head holds sponges and brushes where they can dry between uses. (4th Estate, Paper Chase Farms, Middleburg, Va. Designed by Upperville Barns) Photo by M. F. Harcourt

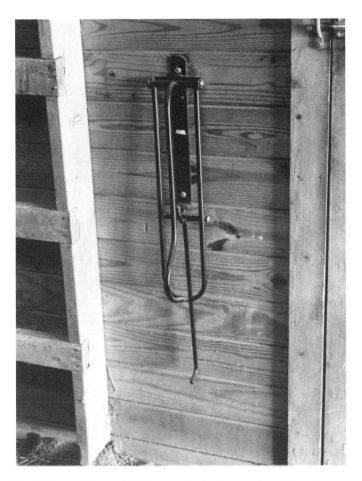

Saddle racks outside each stall can be simple 2″ × 4″s with a hook at one end, or they can be handsome metal racks. Both types fold out of the way when not in use. Photo by N. W. Ambrosiano

If you split a stall space into several storage areas, let one of them be an open aisle like this, with a cabinet at the end. Lined with cedar-chip board and made with widely spaced shelves, this cabinet was designed for moth-proof blanket storage, but holds much more. Plenty of hooks on the walls ensure a tidy place for all sorts of equipment, where it is out of the barn aisle but still accessible. Photo by N. W. Ambrosiano

WASH STALLS

A wash stall is a bit of a luxury for the casual horse owner, but for anyone involved in hard training, showing or breeding, it's a necessity. The term "wash stall" is actually a misnomer. What is really needed is a spot with good drainage, hot and cold water, and cross ties. That can include aisleways and corners of the barnyard, if simplicity is to your liking.

Drainage is the first equipment, as it takes a lot of water to wash a horse. Once it comes off the animal the water must be drained out of the traffic areas. If you choose to install a subfloor drain in a concrete pad, be aware that clogging will be your main nuisance. Make sure the grating over the drain is sturdy, removable, and is cleaned frequently.

Another method is to slope your footing to one wall, and run a 3-inch pipe from the lowest point out through the wall. On the outside, don't hope for nature to handle the runoff, or you'll flood your barn's footings. Either extend the pipe away from the barn or drain it into a drywell filled with gravel.

One northern Virginia eventing stable, that of U.S. Equestrian Team rider Phyllis Dawson, went with the concrete floor option, but added a foot-soaking area for her hard-working horses. Along one wall is a 6′ × 4′ soaking pit approximately 4 to 6 inches deep. The rest of the stall drains into this pit, and the pit drains into a wall pipe as described above, but there is a plug to block the wall pipe during treatments. Horses who won't tolerate a foot in a bucket will stand happily for warm- or cold-water treatments in this open bath.

Concrete is not essential for a floor; gravel over a good base will do, too. Or, if you wash in the barnyard, a good thick patch of grass that drains well makes a fine footing. Just be sure you won't be standing in a puddle or a mud hole by the end of the bath.

You can do without hot water, especially if you're in a warm climate, but it makes frequent bathing and other barn chores more pleasant when the mud is thick on the horse and a chill is in the air.

You'll be happiest if you make your tie space a cross tie or next to a wall so that your horse can't roam around the tie ring during the bath. Opinions differ over where to place tie rings, at wither height or above. In any case, the rings should not be any lower than your horse's withers. He could pull back and damage his neck if tied too low.

The ropes or straps you choose for tying can be anything that allows a quick release in case of emergency. Some people prefer chains, but a horse throwing his head can toss the chain into your face, which is sure to make you a devotee of soft cotton ties. Ties that are too loose are a liability, so measure them carefully. They are about right if they just reach to the

This enclosed wash stall has one unusual feature: a dropped area for foot soaking those horses who will not tolerate buckets. The entire stall drains to this 4′ × 6′ well, which itself drains through a 4-inch pipe in the wall to an outside drainfield. The pipe can be plugged for a soaking session or left open, as shown, so a wet blanket can drip dry on a convenient chain. Photo by N. W. Ambrosiano

These wash stall details are well-arranged here. At left, a nylon cross-tie is attached with a quick-release clip to a screw eye in the wall. Beside it, a hose is safely coiled over hot-and-cold water taps, which are out of reach of a horse's legs. To the right, the essential scraper, hoof picks and brushes hang neatly off the floor, along with an emergency switch. The wall switch for the overhead light is safely enclosed in an exterior-type housing, and the covered plug allows the area to be a clipping stall once the water is drained away. Photo by N. W. Ambrosiano

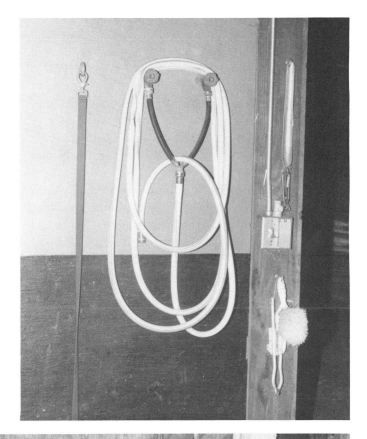

This concrete-floored wash stall has everything close at hand, yet out of a horse's reach. The diagonal shelves at the rear of the stall have no protruding corners to hurt a horse, and the thick cotton cross ties ensure that he stays at the front of the wash area. Drainage could be slightly improved, though. Notice that the water pooling in the corner has discolored the walls. Pressure-treated plywood is a must, even where the walls do not touch the outside soil. Photo by N. W. Ambrosiano

horse's halter from either side. Any looser and the advantage of cross tying in the first place is lost. The horse will be able to turn around or even get a front leg over low ties.

Two schools of thought govern the attachment of ties to the wall. Some prefer strong ties bolted firmly to the wall, with a quick-release catch at the halter. A panicked horse can be released if you deem it necessary and can get to his head, but he can't idly pull himself free.

Others choose to link the ties to wall rings with a loop of baling twine, letting the horse break the twine before hurting himself. Your management style and the temperament of your horses will define your best course here. If you're not sure and you'd like to avoid replacing halters, go with the baling twine option.

If you plan an enclosed wash pit, a fairly standard size is 10' × 10'. This gives you enough room to work around your horse without him crushing you against the wall. If you plan a barn with a stall or two more than you have resident horses, you've got the beginnings of a wash pit right there. If you're sharing space with a storage area or enclosing the end of the aisle, just be sure to give yourself enough room for a full-sized horse and you. Some horses are claustrophobic in a small space, and you need to have enough room to maneuver in case of trouble. If a narrow space is all you have available, make sure you can at least turn a horse around in it without scraping hips and sides.

Since there is more to giving a bath than merely hose and horse, equipment storage is a big help. You need plenty of space for wet sponges and brushes, a place other than underfoot for the scraper, and a reel or rack for the hose. Hanging metal mesh baskets make good gear keepers, as do plastic milk crates. Wooden boxes tend to become waterlogged and rot, and things in the bottom can mold instead of drying. With a little scrap lumber, you can build a set of open corner shelves in the back corner of your wash stall that's deep enough to hold a few things and reinforced with a piece of molding across the front to prevent the contents from rolling out.

Keep your hose rolled securely and out from under the horse's feet to ensure both its longevity and yours—a tangled horse or groom is counterproductive. Just be sure that if you install your hose reel or hanging rack along a wall, your horse cannot hit himself on protruding parts.

Good light is essential in the wash area, and it is even more important here that it be out of reach of a rearing horse. Bathing and associated care are more likely to bring on violent reactions than most other kinds of handling.

HIGH TECH AND SAFETY FEATURES

There are many new high-tech innovations that you can add to your barn if you feel it necessary.

Fire

The worst nightmare of every barnowner is fire. It starts suddenly, drives normally staid horses into a suicidal panic, and sweeps a barn before you can begin to stop it. To keep it as unlikely as possible, you must not only build in safety measures but practice good management.

Sprinkler systems can be easily incorporated into the overall barn construction. This option will pay for itself in insurance savings if you have a public-access stable. Building with fire-resistant materials or specially treated lumber can also add to your peace of mind. Ask your local building supply store what they have available, and check as well with the fire department for its recommendations.

Make sure your doorways are open and accessible in case you need to evacuate horses quickly. All the latches should be operable with one hand. If you plan on boarding more than 10 horses in the barn at once, be sure at least two clear exits are available at all times.

Don't skimp on fire extinguishers. You don't have to have the big, unwieldy institutional ones, especially if you expect small people to use them. Just make sure the ones you get are fully charged and in reachable places that everyone in the barn knows about.

No matter how good a housekeeper you are, take another look around the barn. There should be no hanging, dusty cobwebs that could flame up and no loose piles of hay or bedding against the walls. For the safest arrangement, store bedding in a separate shed. If you have floor-level storage or even a loft, consider lining its floor, walls and ceiling with fire-

Metal connector plates like these at the top of the posts that join two perpendicular 6″ × 6″s can add to the building's expense, but save time and trouble in construction. Photo by N. W. Ambrosiano

Joist hangers are the most secure way to handle such large sections of joists and beams where simple toenailing of perpendicular boards would not be enough. Photo by N.W. Ambrosiano

Finishing the roof as soon as possible gives you a sheltered area to work under, which is invaluable in wet climates or if you build during the winter. (Joe Ann Scott, Fairfax Station, Va.) Photo by N.W. Ambrosiano

retardant materials that will hold a fire back for at least an hour. While a hayloft is not generally recommended, one advantage of it is that fire burns upward. Hopefully, that will give you time to get the horses out before beams begin falling into your path.

A hose with a good sprayer head located at the midpoint of the barn can be a lifesaver. Even if you have automatic waterers and a wash stall at one end, go ahead and add this extra faucet to the plans.

And don't minimize the importance of a phone. Calling the fire department is the first thing to do. No matter what, you'll need to give them a jump on the flames.

Security

There are numerous security systems that will allow you to sleep peacefully at night, safe in the knowledge that your horse has not been taken for ransom or sold to the nearest meat market. In addition to horse theft, stealing tack and related equipment is big business because such materials are so hard to trace and so easy to transport. And no horseman in his

right mind ever passes up a bargain. Just remember the next time you run into one: The shoe could be on the other foot one day.

The cheapest option is a dog with a good bark or, even better, geese or guinea hens. If you haven't been exposed to these organic alarm systems, be assured they can make enough noise to raise the dead. If you have truly valuable animals or a facility where you're responsible for horses and equipment belonging to others, an electronic security system may be necessary.

There may be several reasons why you want to add an electronic security system. It's best to plan accordingly during the early building stages. Check with a reputable company to see if they can devise a system for your barn. Many home systems involve motion detectors, which pose a possible problem if you have dogs, cats, birds or horses milling around at night, so be sure the company understands that. Then, see if you need to plan for any additional electric lines or back-up batteries to accommodate the system being installed.

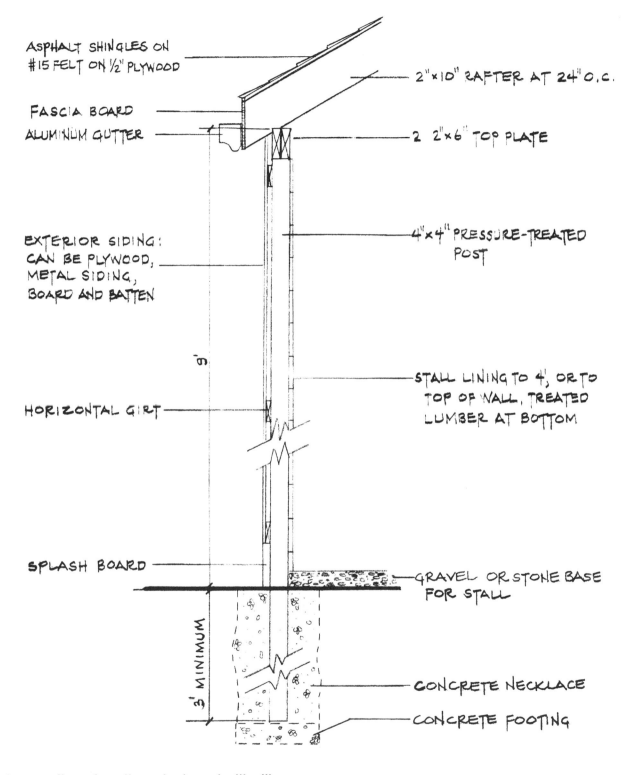

ASPHALT SHINGLES ON #15 FELT ON ½" PLYWOOD

FASCIA BOARD

ALUMINUM GUTTER

EXTERIOR SIDING: CAN BE PLYWOOD, METAL SIDING, BOARD AND BATTEN

HORIZONTAL GIRT

SPLASH BOARD

2"x10" RAFTER AT 24" O.C.

2 2"x6" TOP PLATE

4"x4" PRESSURE-TREATED POST

STALL LINING TO 4', OR TO TOP OF WALL, TREATED LUMBER AT BOTTOM

GRAVEL OR STONE BASE FOR STALL

CONCRETE NECKLACE

CONCRETE FOOTING

9'

3' MINIMUM

This is a section elevation of a barn built with a modified pole construction. Rather than eave girts bolted to the poles, a top plate is used on which the rafters rest.

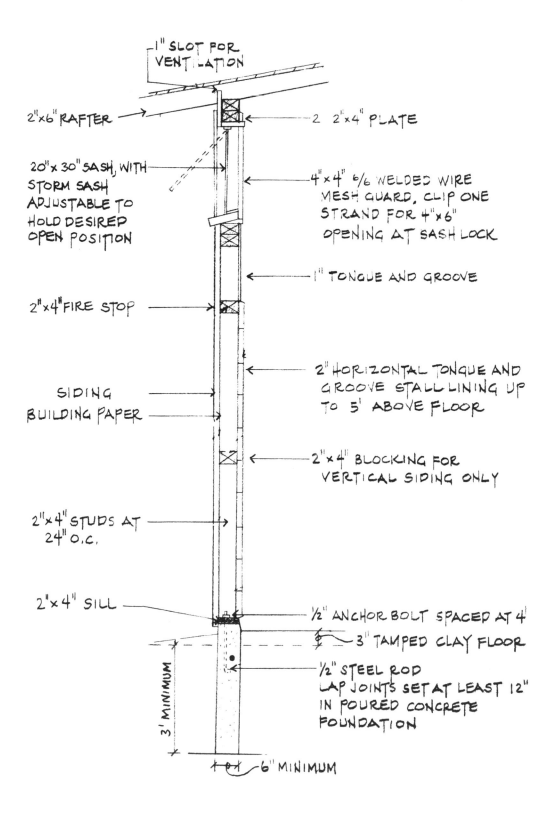

In a barn built on a traditional foundation, here
is a section elevation of the wall which includes
a window with a wire mesh cover.

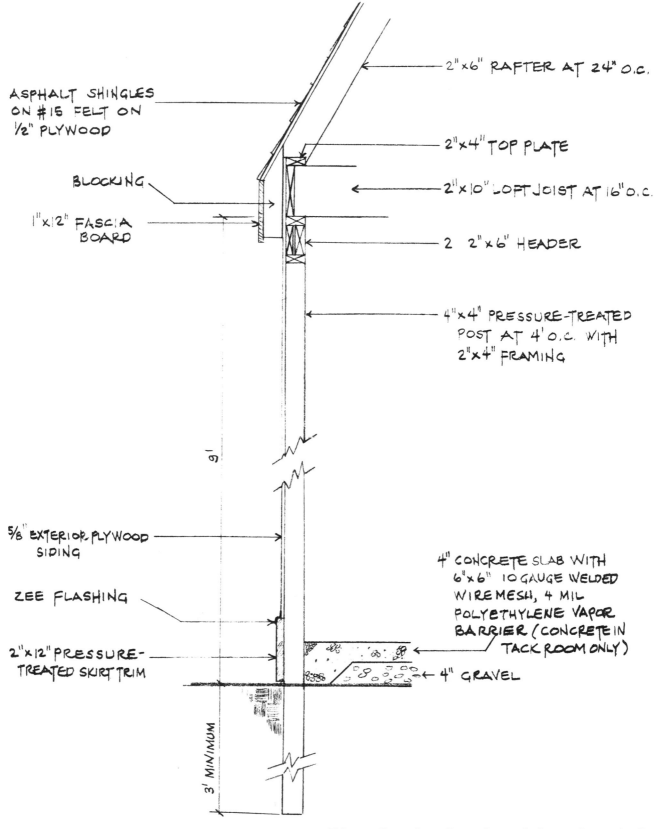

ASPHALT SHINGLES ON #15 FELT ON ½" PLYWOOD

BLOCKING

1"x12" FASCIA BOARD

2"x6" RAFTER AT 24" O.C.

2"x4" TOP PLATE

2"x10" LOFT JOIST AT 16" O.C.

2 2"x6" HEADER

4"x4" PRESSURE-TREATED POST AT 4' O.C. WITH 2"x4" FRAMING

9'

⅝" EXTERIOR PLYWOOD SIDING

ZEE FLASHING

4" CONCRETE SLAB WITH 6"x6" 10 GAUGE WELDED WIRE MESH, 4 MIL POLYETHYLENE VAPOR BARRIER (CONCRETE IN TACK ROOM ONLY)

2"x12" PRESSURE-TREATED SKIRT TRIM

4" GRAVEL

3' MINIMUM

This section elevation of a pole barn shows both the flooring for a tack room and the support detail for a steeply pitched gambrel roof.

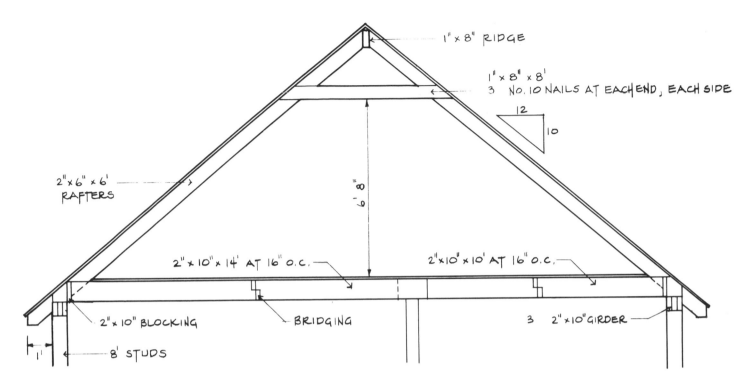

A simple section of a gable roof includes a collar tie support on the rafters. Rafters are shown as 2″ × 6″s but with wider barns or ones with large snow loads use 2″ × 8″s.

Intercoms

Intercoms can be part of your security system and can also let you monitor ongoing activities in the stable. For example, they are quite handy for checking on mares about to foal or colicky horses or for catching the horse that has a tendency to cast itself. Of course, an intercom also makes it easier for you to keep track of the kids, call folks for dinner, or whatever.

Intercoms are very simple to purchase and install yourself or you can have them installed for you. Your pocketbook and needs will determine your choice. For as little as $30 you can install a system usually sold as a nursery monitor that will allow you to keep an ear in the barn, even though it's only a one-way device. Two-way systems are easily available as well, and you can even install a complete loudspeaker system that allows you to use your facility for horse shows. You can find intercoms at consumer electronic stores or check catalogs for special packages.

A Phone in the Barn

A phone in the barn is an absolute necessity these days, even if it seems a bit frivolous when you're there to ride, not talk. In these times of car phones and phones on airplanes, why not have one in the barn? It can be a venerable rotary dial or a cordless phone that can go into the arena with you. If you use the barn as a refuge from the phone, you can always turn the bell off. After all, it's outgoing emergency calls you really need it for. Just for the sake of fires, veterinarians and general communications, hook one up.

Closed-Circuit Television

Hearing broodmares about to foal is a little tricky, but seeing what they are doing on closed-circuit television leaves no doubt. Here again, this piece of technology has its place in the right type of facility and may be incorporated as part of the security system.

Closed-circuit television systems are expensive if you have a small-scale operation. Plan on at least $3,300 for a one-camera setup. But if you have good mares bred to expensive stallions, you might consider this as an option to hanging out in the barn all night waiting for a sneaky mare to foal.

6 COSTS

Figuring costs for your barn isn't hard as long as you deal with ballpark estimates. You can usually build a barn yourself for half the price of contracting it out, provided you count your time and labor as being worth nothing. If your needs are simple and you're handy as a builder, that is clearly the way to go, especially if you take advantage of such prefabricated items as sliding door kits and stock-size trusses.

Most horse owners with more than a run-in shed in mind enlist the services of a barn builder who may or may not have a selection of his own plans to offer. Most builders will happily tackle any set of plans, of course, provided they are familiar with the requirements and can get the needed materials. If you plan to use your house builder for the barn, be aware that his unfamiliarity with barn methods may lead to a higher estimate than otherwise, because he's working from a different perspective than an experienced agricultural builder.

In general, you might assume that a barn's cost per square foot, as executed by a barn builder, will be under one quarter of the cost of an average home. Quotes in 1988 for East Coast barns ran approximately $8/sq ft for a simple shed-row barn, $10/sq ft for a center-aisle barn, and $12/sq ft for a simple center-aisle barn with overhead hay storage. Add a few customizing touches and you're approaching $20/sq ft.

Depending on availability in your area, you may be able to save a lot of money simply by good planning. Taking steps like building in 12-foot increments can allow you to buy all your lumber in that size, which is standard, stocked dimension and is often sold at a better price per foot than other sizes. Many distributors stock trusses in a 24-foot width at approximately $40 each, making that an economical choice if you plan to store hay elsewhere.

Remember that lumber stores like to deal in volume. If you call for prices on sizes of lumber, be sure to explain the amounts you're dealing with, and see if they will cut you a better deal. If you run short of cash, concentrate on getting the full exterior built first and make do on the inside. Solid doors, hay racks and watering systems can come much later, and your money is better spent on a good price for plenty of pressure-treated lumber.

If you wish to get the most barn for the money, remember that a long barn is cheaper than a wide one. Increasing your roof span will cost more than adding a truss or two on the end. If you want larger stalls than a plan calls for, you can shrink a center aisle to 10 feet, but no smaller—you won't be able to safely turn a horse around or pass by a vehicle parked indoors.

When thinking of overhangs—those lovely items that provide shade in the summer and shelter in the winter—figure that on a 36-foot barn, an average front overhang will cost $1,500. Multiply or divide that figure as needed for your plan to see if you can still swing it.

If building materials in your area are expensive, look into the prefabricated market. There are many reputable companies advertising nationwide that can work with you to provide a shell for you to fill or fill the shell you provide. Even with shipping costs, they can make a competitive bid in many cases.

Choose materials that are easy to find in your area. For example a set of plans may call for 4″ × 4″ posts, but you have similarly sized, pressure-treated poles nearby at a good price. Don't hesitate to substitute materials, provided the new ones are at least as strong, weatherproof and nontoxic as the ones called for.

7 CONSTRUCTION

STEP-BY-STEP PLANNING AND PROCEDURES

Let's assume that you have selected your land and now you need a step-by-step list of things to do to head you in the right direction.

1. Check out the zoning regulations, municipal building codes, covenant restrictions and any other stipulations that may affect your plans. Often there are also regulations concerning land use/water sheds and pollution from soil runoff that you need to investigate.
   - Suggested sources for this information:
   - Municipal zoning/planning offices
   - Building Inspector's Office
   - Soil Conservation Service
2. Make appointments at the appropriate information offices, and establish a positive relationship with these people. A run-in with an inspector can sour your whole project and delay every step. Ask the inspector staff for their advice on your project, and take it. While they're probably not horsepeople, they know the local building scene and can steer you away from potential problems if they so choose.

   Set up a file system so that you have easy access to the information you gather. Departments and service offices available to you are:
   - Cooperative Extension Service
   - Well-respected local horse owners
   - Equine consultants
   - Soil Conservation Service
   - Forestry Service (If your land is wooded and you need information on getting the timber removed, and so forth.)
   - Agricultural Stabilization and Conservation Service

3. Begin networking with the local horse community by asking about groups, organizations and educational meetings available at local feed and tack shops. Get on their mailing lists.
4. Subscribe to one or two quality horse magazines that give you current information on horses and horse management. (As your experience level increases or your style of riding or direction in the industry changes, you may want to drop some and subscribe to others. Be flexible!)
5. Begin asking yourself what you want, need and have to have for your operation. If your budget is limited, then plan a step-by-step list of what's essential to start your operation and what can be added later.
6. Locate a barn builder in your area who has a good reputation for quality work and reliability. Look at several barns he has built and talk to the owners. Never hesitate to ask for references or referrals on your builder. While there may be many well-qualified builders in your area, we strongly recommend someone who has built horse barns and/or is a horseman himself.
7. If you plan on being your own builder, locate some subcontractors unless you are confident about your wiring, concrete-laying and woodworking abilities. Building the facility, if it is on a small scale, is not difficult. However, if you are not 100 percent sure about the electrical and plumbing aspect of your barn, you might do better to have them subcontracted.

   Even if you are planning on being your own builder, you may find it worthwhile to pay a local barn builder for some of his knowledge, using him on a consultant basis. His knowledge of local building codes, soil problems and material availability is invaluable.

8. Locate and price local firms that sell lumber and other building supplies you need. Here again, begin networking. Establish an account and credit with a good company so that you can plan a budget for your building schedule. The store can also be a good source of how-to information as well as a source of suggestions for materials and substitute items that you will need.

9. Invite any local SCS or Extension personnel who will make an on-site visit to come and make recommendations as to location of facilities, angle of placement due to prevailing winds, suitability of soil and so on. Then lay out your plans on paper to get an idea of the movement of horses, machinery and pedestrians.

10. Think about drainage before you put down the foundation. If you're going to build up your barn site, investigate local materials that will be economical to provide the dry flooring you need in the barn. Will your current fill material be suitable, or will you want to add drain lines in the stalls? Plan to dig these before adding the fill. Put in any drain lines you need for the wash pit, washer/dryer and bathroom if you plan on adding them at any time. Remember that you don't have to put them in right now, but if you haven't made arrangements for them it will be both more expensive and more trouble to do so later.

11. Now go to the building order list that follows and begin construction. Monitor your progress daily and be ready to change plans if necessity demands. Have a scheduled completion date, and if you're using a builder, work with him to meet the building deadline. If building it yourself, you will know why if you get off schedule.

Building Order List

Assuming that you have selected the construction site for your ideal barn, here are some procedural steps to aid you in getting things accomplished in timely and logical order.

1. Site selection based on:
   - Drainage
   - Prevailing winds
   - Accessibility to roads, rails, airports or whatever forms of transportation will be necessary to your operation
   - Availability of water, utilities
   - Passive/active solar considerations
   - Future improvements, additions
     —Arenas, riding trails
     —Parking
     —Fencing
     —Ponds
     —Facility expansion

2. Site preparation
   - Excavating, grading for roadways, drainage and so on
   - Preliminary utilities layout
   - Building permits
   - Construction utilities installation (electric, water)

3. Construction
   - Pole Barn
     —Mark exterior building lines, set batter boards, excavation stakes, and so on
     —Dig drainage trenches
     —Mark pole locations, dig holes
     —Set poles, pour concrete necklaces
     —Backfill holes, flood after 24 hours to settle fill
     —Attach and level splash board, girts
     —Add additional fill for stall drainage
     —Begin roof work
       - Set eave girts, or, if resting a top plate on leveled pole tops, attach it
       - Set rafter girt, if used
       - Step off rafters for gable roof and cut lines where rafters will meet plate or eave girts. If using prefabricated trusses, strap them in place with temporary framing
       - Attach collar ties to rafters
       - Chalk line along tails of rafters; cut even for fascia board attachment
       - Frame eaves with fascia boards and soffits, leaving space for vents
       - Deck and cover roof with chosen material
     —If planning a hayloft, lay joists on eave girts or suspend them from girts with metal joist hangers

—Lay plywood loft floor

—Place flooring in tack rooms, feed areas

—Nail interior barn walls (rough-cut 1″ × 6″ or 2″ × 6″ boards) in stalls, frame for windows, paneling, water lines, and so on in tack room

—Install windows, doors

—Add exterior siding

—Stain, add gutters, finishing touches

• Conventional foundation structure (This assumes you are building a full house. Delete any steps not applicable to your barn. Knowledge of local inspection requirements is necessary.)

—Set foundation—batten boards, excavation stakes, and so on

—Dig footers

   • Form footers and reinforce rebar—add drainage aggregate

   • Pour footers, remove forms

—Alternate For Poured Walls

   • Set wall forms, plumb and level-add reinforcement, rebar-add drainage aggregate

   • Pour walls and remove forms

—Install top plate

—Add center stringers if required

—Construct first floor joist

—Build subfloor

—Any time after #6, French drains may be installed and foundation backfilled

—Erect exterior/interior wall studs

—Install second floor joist, if applicable

—Lay second floor subfloor, if applicable

—Add top plate

—Install prefabricated trusses or stick-build rafter system

—Erect roof sheathing

—Construct roof felt and roof

—Install exterior wall sheathing, insulation board, vapor barriers and so on

—Rough-in plumbing, electrical work

At some time during the previous steps, concurrent provisions should be made for drilling well or obtaining other sources of water; electrical services should be connected to a service entrance; and septic system should be installed or provisions made for connection to existing sewage disposal system

—Begin exterior siding, masonry

—Add drywall installation

HOW TO'S

Laying out Your Barn

Decide which way you want your barn to face, based on your land and the prevailing winds. Then note the exact length of the outer edge of each wall by placing a peg where one wall will begin, measure to the other end of the wall, and drive another peg. Run a string from a nail in the center of the top of each peg to mark off a straight line along the wall.

To line up the other walls, use the 3-4-5 rule from your old geometry days: To form a right triangle, use those dimensions, or multiples thereof, and the right angle will be in the corner opposite the 5-foot side. To be sure of your measurements, multiply this out to at least 6 to 10 feet, and be sure to measure all corners this way. If your first markings are square, it will make life much easier and save you money on extra lumber when you get to the roof!

Erecting batter boards, once you have square corners, allows you to remove your string lines to set poles, then replace them without going off-track along the way. In each corner, just outside the outer

framing line of the barn, sink three pegs conforming to the corner. Tack two boards along the pegs, and put a nail along the top of each board to attach the string exactly along the foundation lines. Replacing the string accurately after each removal will be no problem with the batter boards in place.

Pole Buildings

A pole building is so called because it is built around poles or posts that are set in the ground. Rather than sitting on a foundation, the barn in effect hangs from a frame of pressure-treated lumber. Spare no expense on your main posts, and be sure to check with local soil experts on the depth of hole you'll need for a solid placement. If you have brought in fill dirt, you may be required by local codes to dig down to undisturbed soil with each post, adding to your lumber expense. Alternately, you may choose to surround each post with a "necklace" of concrete poured into the hole for security.

Before digging the post holes, mark the location for each along the string line. As the lines are marking the outer edge of the barn, you will need to locate the holes by measuring the width of the siding girt plus the center measure of your pole; allow 2 inches for the girt plus 3 inches for a 6-inch diameter pole. Accurate marking will save you from constructing walls that "wander" slightly and throw off your measurements for such standard items as plywood panels.

If you choose to build using 4" × 4"s, a handy spacing for them is every 4 feet, allowing room for doorways and windows and plenty of support for the roof and loft. With posts 4 feet on center (o.c.), you always have a firm support for both hinges and latch on every door you install. Be sure, though, that if you make or order doors and plan to set them inside these posts, you take the 4-inch smaller inside dimension into account.

Using 6- and 8-foot pole increments is also effective if you use larger poles or posts. These match standard units in which lumber and siding are sold, which will save you milling costs.

Once the poles are in place, the splash board goes on, that is, the board that skirts the entire bottom edge of the building. Also known as the sill girt, it is the first of the girts that supports all of your siding, walls and rafters for the building. Use a 2" × 6" or 2" × 8" pressure-treated board here. If this first splash board is placed level, it represents a true horizontal for the rest of the building, and you can line up your siding and other work from this level board. You can pack stall flooring in any time after this treated board is in place.

With the poles and splash boards up, you can go directly to the roofing, if you'd like shelter as you complete the walls and interior, or you can work from the ground up, framing in the stalls. See the Building Order List.

Exteriors

One of the most popular exterior sidings for modern barns is T1-11 plywood, an exterior grade of ply that has a reverse board-and-batten appearance. It can be stained to match nearby buildings and comes in standard 4' × 8' dimensions. Install it vertically, and don't bother to cut out window holes before hanging it. It's easier to simply trim out the windows through the installed sheet with a hand-held power saw.

To join vertical sections of plywood, use special pre-made metal flashing if you want a weather-proof seal. Better yet, use a plexiglass panel for the top 2 feet of a 10-foot wall, adding to your inside lighting.

True board and batten makes a labor-intensive but handsome exterior. The boards are 1" × 6" or 1" × 8" lumber nailed vertically to the siding girts with narrow, 1 or 2-inch, batten strips tacked over the cracks between them. As with any vertical siding, you must first apply horizontal nailers between the posts on which to tack the siding.

Tongue-in-groove, or shiplap, siding is another weather-tight siding that, if installed with the boards horizontal, lets you replace the ground level board as it weathers. Vertically placed sidings that rot at ground level must be replaced entirely, or patched along the bottom.

To avoid having to repaint every year or so, use wood stain in your color of choice on plywood or board siding. Oil stains last longer than paint, and, while they weather after a few years, they don't require painting or sanding before you apply the next coat.

Metal sidings are popular in some areas, and can be extremely cheap and weather resistant. However, they do dent easily, so they must be securely lined against stray hooves and hips.

These metal sidings come in 8-foot lengths and 32-inch widths, but they can be special-ordered in any size you desire. Special fasteners—nails with washers built-in to prevent rattling and leakage—are required. Check the type of nail with the siding you buy, as you cannot use aluminum siding with galvanized nails. You can buy metal siding in a wide range of colors, which allows you to escape a large project —that of painting what you've built.

Roofs

You have three main options in your roof construction: large timber beams with lumber decking, prefabricated wood trusses or a stick-built system. Local material costs and availability will have a strong bearing on the type you choose, but so will the amount of storage you wish overhead. For the clearest storage space, timber beams, supported if necessary by a horizontal beam between the base and the ridgepole, are your best choice.

The most economical roofing system is often prefabricated trusses. These can be purchased from a fabricator, who will make them to fit your specifications. They preclude any overhead storage because of their many cross webs and the fact that they are usually installed 24 inches on center.

House builders will often plan a stick-built system for your barn, which offers more storage capacity upstairs as it eliminates the diagonal webbs needed in trusses.

No matter which roof system you choose, plan a 6- to 8-inch open space beneath the eaves to allow for good overhead ventilation. Should you wish to block this off occasionally, such as during a storm, you can install a hinged board at the top of the wall that can be swung up to close the gap.

Roofing Materials

You have a wide range of roofing material options, depending on the appearance you desire. In the Southwest, handsome clay tiles make a cool, waterproof roof, blending well with the local construction style. In Eastern suburban neighborhoods, asphalt shingles make the barn fit in well and provide a good, durable surface.

For the cheapest roof, metal, whether corrugated steel or aluminum, is your answer. It can be noisy in rainstorms, but it is extremely easy to install and requires no maintenance. If you live on the coast, where salt breezes wreak havoc on regular steel roofs, aluminum is your best choice. Remember to use only aluminum nails with this, or the roof will react chemically and become damaged.

A metal roof is installed over a simple grid of nailers between the rafters, giving you the fastest shelter with the fewest materials. It requires no heavy decking or underlayment and has a long lifespan.

Durable asphalt shingles will last up to 20 years and can be ordered in colors that complement nearby buildings. However, they do require considerable preparation: You must lay a ½-inch plywood decking over the rafters, then a layer of roofing felt and finally the shingles themselves. Instructions for installation vary with the brand of shingles, but they are usually included on the package.

Regardless of the roofing material, plan for gutters on your barn. Tacked on to the fascia boards on the ends of the rafters, they will keep your barn from disappearing behind a sheet of water in every rain storm. Vinyl or aluminum gutters are inexpensive, simple to install, and adequate in all areas but those of high snowfall. If you have a steeply pitched roof and severe snow, ask local building specialists for their recommendations. You may need to resort to reinforced guttering, brackets to hold back snow, or other options.

If you choose to go without gutters, you can run a grade strip of gravel away from the barn that will catch the falling water. First, dig a trench along the drip line, leading to a drain field, and then fill it with gravel and edge it with scrap lumber.

8 BARN PLANS

FLORIDA SUN SHED

Description.

A run-in shed is the ideal arrangement for horses in all but the most severe climates. The animals are allowed shelter, but not enclosed where they can develop leg and lung problems. In Florida, as in many parts of the country, a shed need only offer shade from the sun—wind and snow are not encountered. Thus, this shed at Hassel Arabian Stud in Reddick, Fla. has no pitch to its roof, serving as little more than a permanent parasol and hay feeding area.

Materials

Posts:
  6″ × 6″ pressure-treated or creosoted
Rafters:
  2″ × 8″ nailed with 16d nails
Framing:
  2″ × 8″ joists
Roof:
  Sheet metal with nails of same metal

This variation on the run-in shed allows for three sides to be closed for more wind/rain protection and yet the high roof and wide open area on top of the walls allows for heat to rise and winds to dispel it. Here the hay rack is located on the side wall, giving easier access to those refilling it and providing a safe location in which horses can eat. Placed to the rear of a small shed, a hay rack can inspire one horse to keep others at bay with his hindquarters as he eats. If the rack is on the side, as shown here, he is less likely to dominate the entire shed. (Hassell Arabian Stud, Reddick, Fla.) Photo by M. F. Harcourt

A close-up shot shows a hay rack with a trough below it that catches fallen leaves from legume hays and allows for the feeding of salt, minerals, supplements or grain. (Hassell Arabian Stud, Reddick, Fla.) Photo by M. F. Harcourt

Paul Hassell has found a way to provide some protection for his horses from the Florida sun by erecting these simple open run-in sheds. Since the climate is rarely severe, they function as year-round shelter for both his horses and his cattle. Photo by M. F. Harcourt

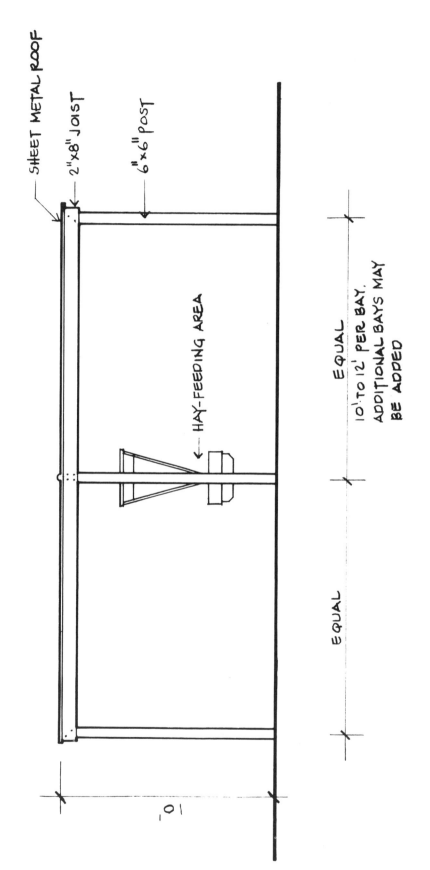

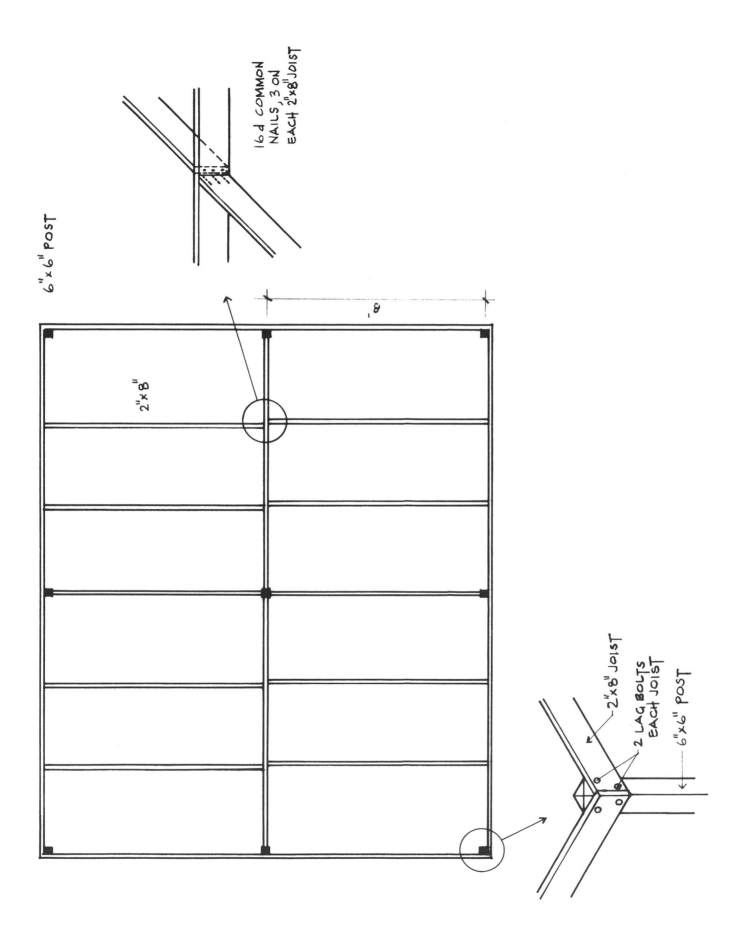

6"x6" POST

16d COMMON
NAILS, 3 ON
EACH 2"x8" JOIST

2"x8"

8'

2"x8" JOIST

2 LAG BOLTS
EACH JOIST

6"x6" POST

CALIFORNIA RUN-IN SHED

Description.

This shed arrangement in Nicasio, California, at Carmen Johnson's Windfield Station, is ideal in an area with paddocks for individual horses, or it can be set over the junction of two full pastures. There is horse-proof storage in the center aisle, and the sides can be enclosed as snugly as your climate requires. To keep a horse in overnight, such as before a show or hunt, one could use a 2″ × 8″ board that drops into place across the main opening. To secure it, make a small channel of 2″ × 2″ at your horse's chest height that each end of the board will drop into. Or, nail a horseshoe to each corner support across the front, with the shoe extending halfway out from the post. The board will slip into the hook provided.

Materials

Posts:
  4″ × 6″ pressure-treated or creosoted
Rafters:
  2″ × 6″ 16″ o.c.
Framing:
  4″ × 6″ top plate
  2″ × 4″ furring for siding
  2″ × 4″ collar ties
  2″ × 4″ sill for shed floor
Siding:
  ⅝-inch plywood
Trim:
  T1-11 plywood facing
Flooring:
  Packed clay on grade in stall area
  Poured concrete slab in storage aisle

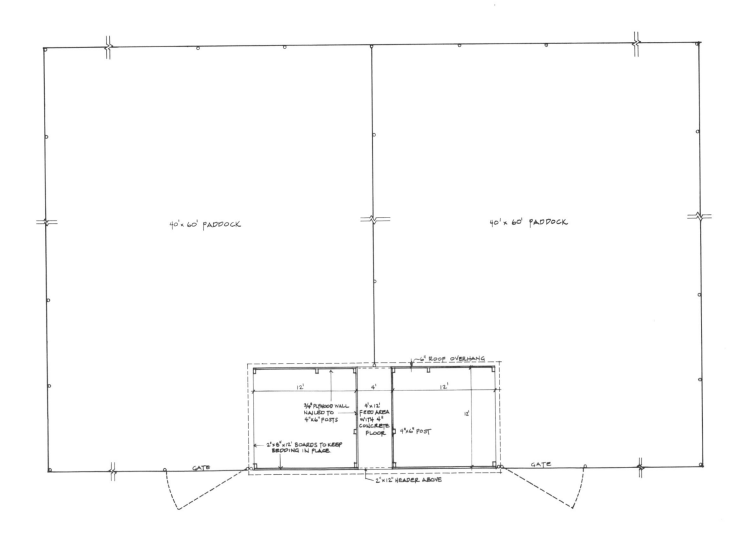

In this California boarding farm, a row of paddocks and run-in sheds allows easy access for each owner and enough shelter for horses in the mild climate. Each shed is centered over the fence between two paddocks, allowing one roof to serve two sheds, with a small storage area between them. Photo by N. W. Ambrosiano

Each 28-foot roof in these shed combinations shelters a 12′ × 12′ area for each horse, plus a 4′ × 12′ storage area between them. A 4″ × 4″-inch beam along the sill of the shelter keeps bedding from being trailed out into the paddock. The support posts are 6″ × 6″ timbers, while the roof is of principal rafter and purlin construction. Fiberglass reinforced plastic or corrugated metal may be used for roofing. Photo by N. W. Ambrosiano

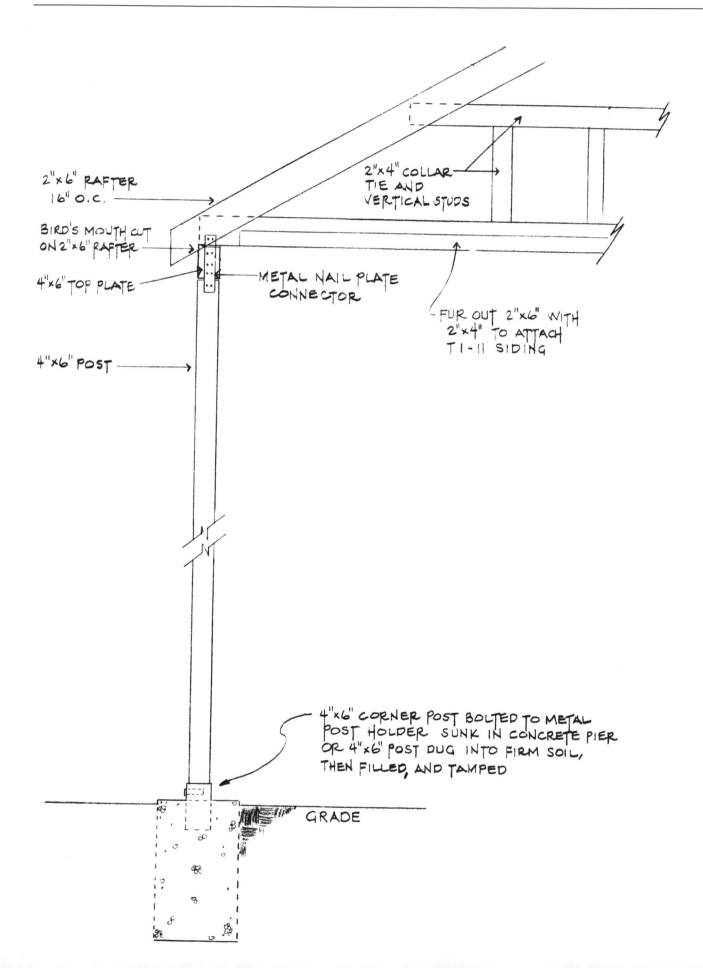

2"x6" RAFTER
16" O.C.

BIRD'S MOUTH CUT
ON 2"x6" RAFTER

4"x6" TOP PLATE

4"x6" POST

2"x4" COLLAR
TIE AND
VERTICAL STUDS

METAL NAIL PLATE
CONNECTOR

FUR OUT 2"x6" WITH
2"x4" TO ATTACH
T1-11 SIDING

4"x6" CORNER POST BOLTED TO METAL
POST HOLDER SUNK IN CONCRETE PIER
OR 4"x6" POST DUG INTO FIRM SOIL,
THEN FILLED, AND TAMPED

GRADE

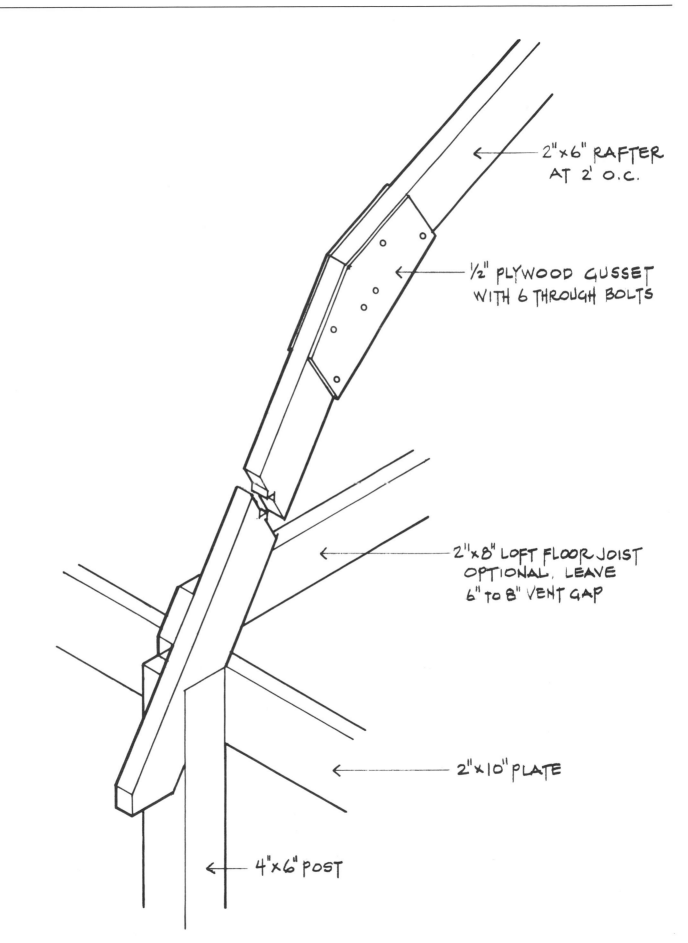

2"x6" RAFTER
AT 2' O.C.

½" PLYWOOD GUSSET
WITH 6 THROUGH BOLTS

2"x8" LOFT FLOOR JOIST
OPTIONAL, LEAVE
6" TO 8" VENT GAP

2"x10" PLATE

4"x6" POST

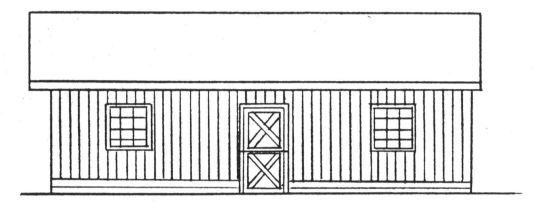

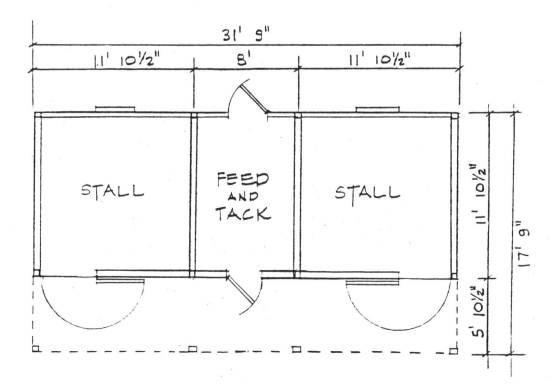

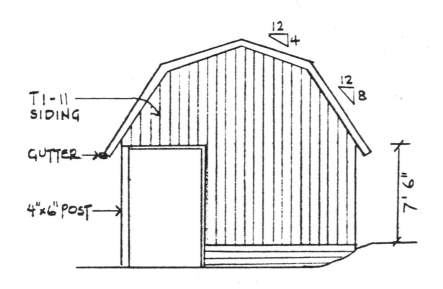

T1-11
SIDING

GUTTER →

4"×6" POST →

7'6"

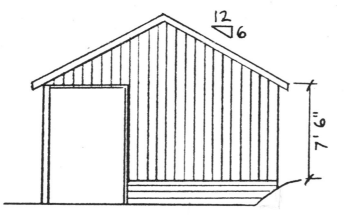

7'6"

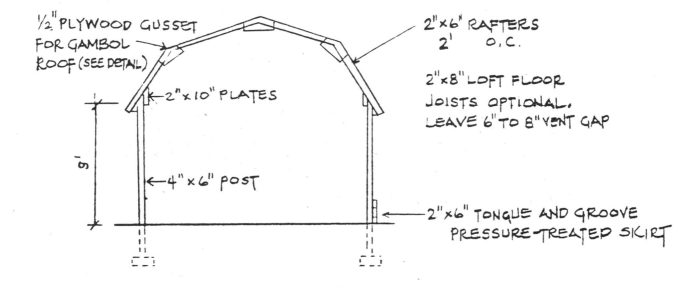

½" PLYWOOD GUSSET
FOR GAMBOL
ROOF (SEE DETAIL) →

← 2"×10" PLATES

9'

← 4"×6" POST

2"×6" RAFTERS
2'   O.C.

2"×8" LOFT FLOOR
JOISTS OPTIONAL.
LEAVE 6" TO 8" VENT GAP

2"×6" TONGUE AND GROOVE
PRESSURE-TREATED SKIRT

3 or 4-Horse Simple Barn

Description.

This simple, workmanlike barn is one that can be built by an inexperienced person with a minimum of trouble. With no overhead hay storage, it provides plenty of airflow through the rafters, its simple pole construction allows for varied ground slope around it, and it can be added to with a minimum of trouble. In its simplest form, it is just stall spaces with an overhang. A 6′ × 8′ storage area can easily be added at the left end of the aisle where it won't interfere with access to any stalls. While this version, built by Helen Makarov of Middleburg, Virginia, is shown with board and batten siding, any other can be substituted with only minor modification of the horizontal nailers between the girts.

Materials

Posts:
  4″ × 4″ pressure-treated (p.t.)
Framing:
  Sill girt 2″ × 12″ p.t.
  Eave girts (2) 2″ × 10″
  Collar-tie girts (2) 2″ × 6″
  Rafters 2″ × 8″ 24″ o.c
  Ridge Plate 2″ × 10″
  Collar ties 2″ × 6″ 48″ o.c. with ply gusset
  Siding girt 2″ × 4″
Roofing:
  corrugated metal on 2″ × 4″ purlins or nailers
Siding:
  1″ × 12″ rough cut oak
  1″ × 3″ battens
Doors:
  Dutch, sliding or as desired

The simplest of arrangements have been made here, such as a central water supply with a hose to reach into the stalls. The plain sheet-metal roof, while noisy in rain, is economical and serviceable.

This simple 20' × 40' barn can be used anywhere you need the basics of shelter, even if your site is uneven and you choose not to fill extensively. Note that in this barn, built by Helen Makarov of Philomont, Va., the barn has been built into the sloping site. While the roofline is level, the stalls are stepped down the hill. As a result, the left and right outside walls are not the same height, resulting in varied material requirements. Photo by M. F. Harcourt

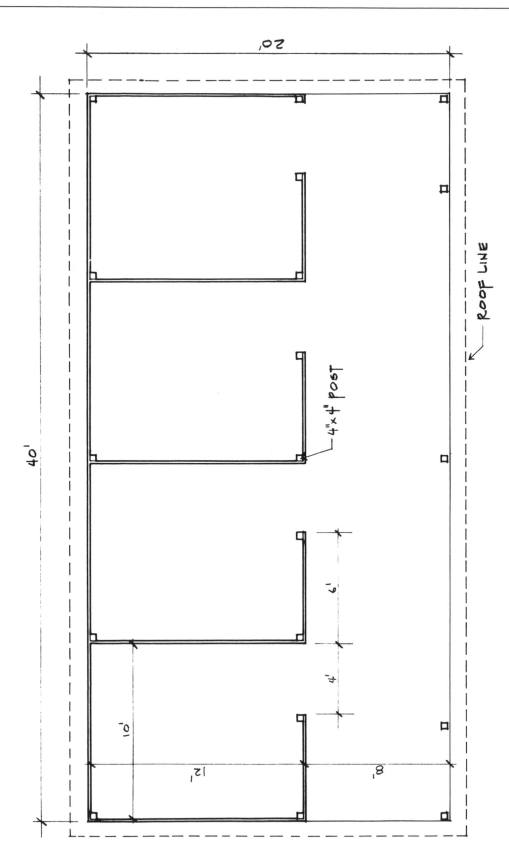

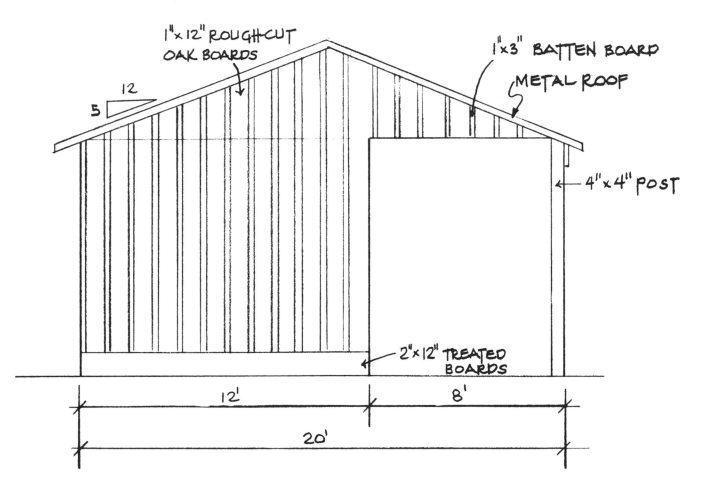

1"x12" ROUGH CUT
OAK BOARDS

1"x3" BATTEN BOARD

METAL ROOF

12

5

4"x4" POST

2"x12" TREATED
BOARDS

12'

8'

20'

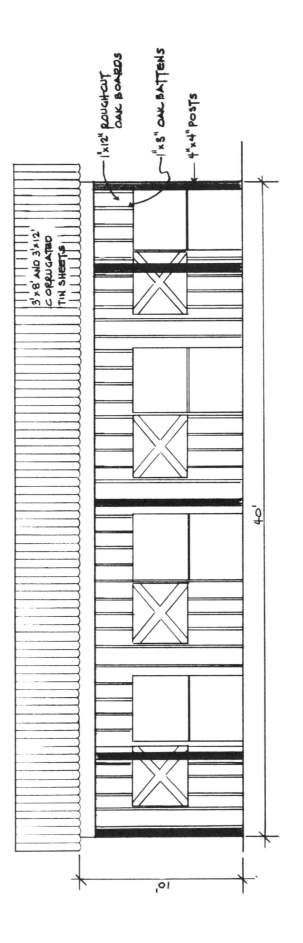

1"x12" ROUGHCUT OAK BOARDS

1"x3" OAK BATTENS

4"x4" POSTS

3'x8' AND 3'x12' CORRUGATED TIN SHEETS

40'

10'

6-Horse Show Barn

Description.

A handsome barn like this is ideal for one or two people to show from. It's moderate in size but has all the details that make preparation for showing a pleasure—wash stall, kitchenette and so on. The stalls' walls are all solid up to 4 feet, then iron rods extend up the rest of the way, giving delightful ventilation year round, especially when windows and doors are open for full cross-ventilation. In addition, the undersides of the aisle loft's 2″ × 6″s were finished with a plywood ceiling, adding elegance and no open surfaces for cobwebs.

Finish it inside as elegantly as you please, or rough in the extras and just enjoy the basic open design's work areas. This was designed and built by the Upperville Barn Co. of Upperville, Virginia.

Materials

Poles:
    20′ class 6 (6″ diameter)
Framing:
    Sill girt 2″ × 12″
    Siding girts 2″ × 6″ set 24″ o.c.
    Rafter girts double 2″ × 12″
    Rafters 2″ × 12″ set 24″ o.c. with 2″ × 12″ ridge plate
    Collar tie girts 2″ × 8″
    Collar ties 2″ × 8″ set 48″ o.c. with ply gusset
Roof:
    ½″ CDX ply
    15-lb roofing felt
    235-lb fiberglass shingles
Siding:
    T1-11
Trim:
    1″ × 8″ fascia
    1″ × 6″ soffit boards
    1″ × 4″ corner boards and casing
Doors:
    Dutch or sliding as desired on stall doors
    Aisle doors double sliders
Windows:
    Closeable shutters only

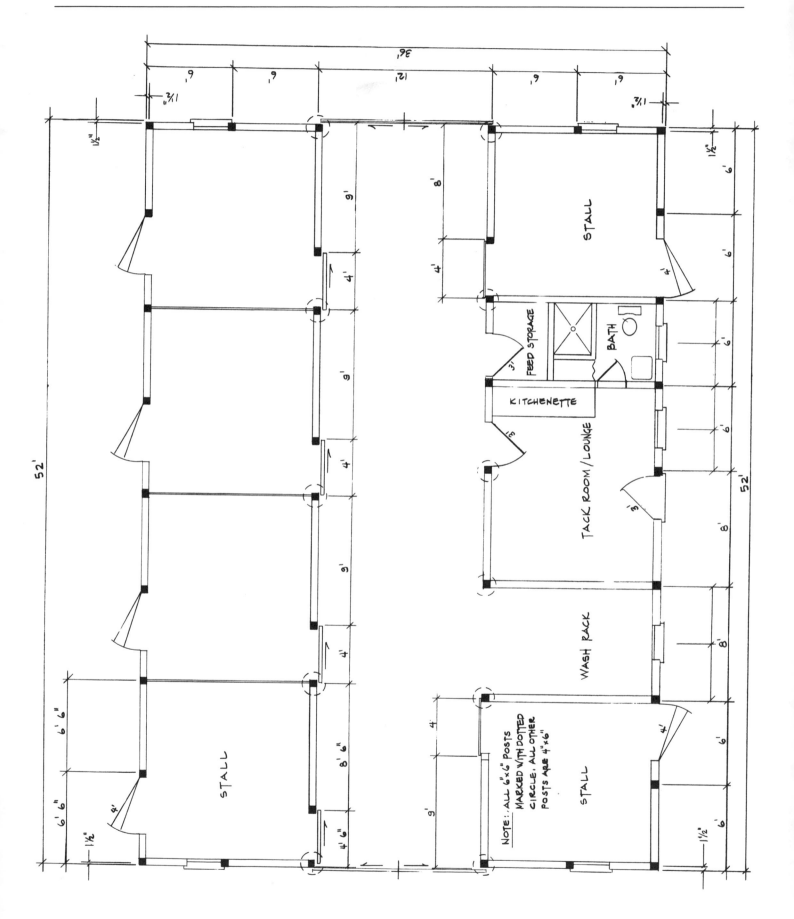

The Fourth Estate barn at Paperchase Farms in Middleburg, VA , was built by Upperville Barns.
Photo by N.W. Ambrosiano.

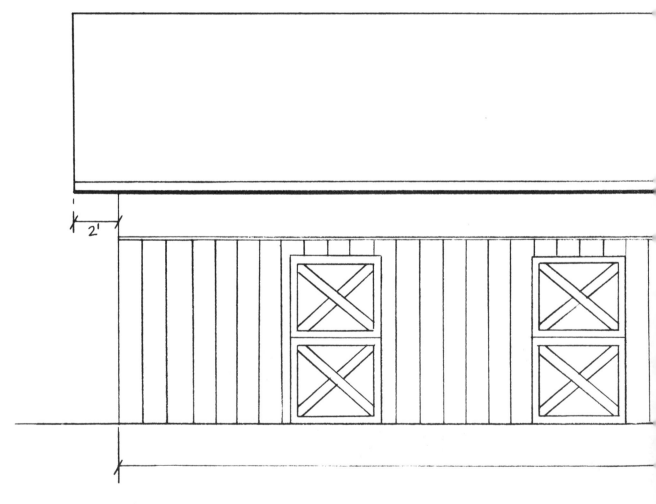

2'

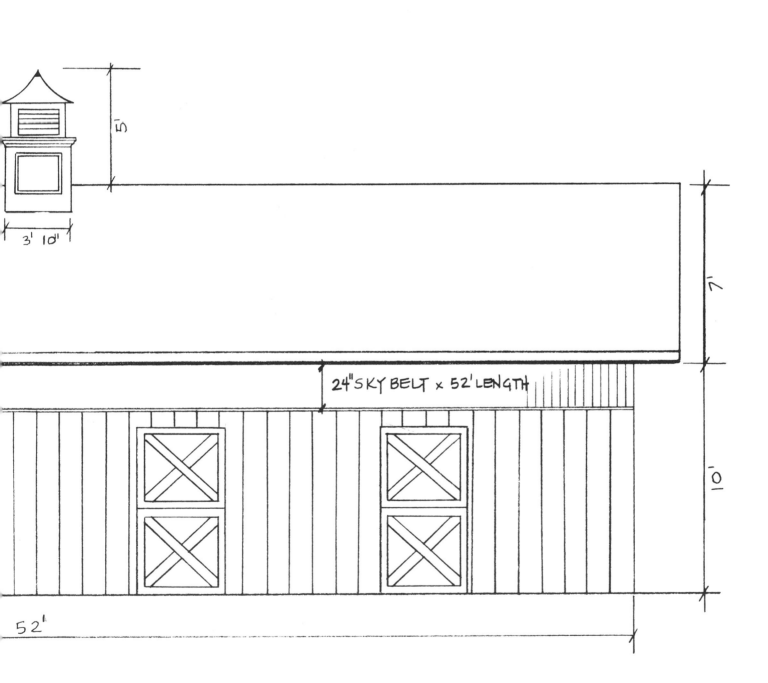

5'

3' 10'

7'

24" SKY BELT x 52' LENGTH

10'

52'

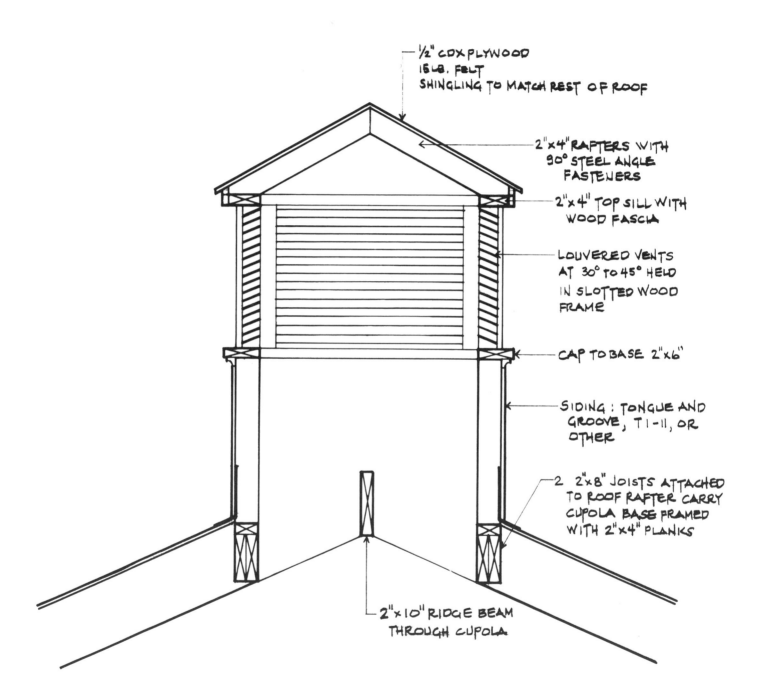

½" CDX PLYWOOD
16 LB. FELT
SHINGLING TO MATCH REST OF ROOF

2"x4" RAFTERS WITH
90° STEEL ANGLE
FASTENERS

2"x4" TOP SILL WITH
WOOD FASCIA

LOUVERED VENTS
AT 30° TO 45° HELD
IN SLOTTED WOOD
FRAME

CAP TO BASE 2"x6"

SIDING: TONGUE AND
GROOVE, T1-11, OR
OTHER

2 2"x8" JOISTS ATTACHED
TO ROOF RAFTER CARRY
CUPOLA BASE FRAMED
WITH 2"x4" PLANKS

2"x10" RIDGE BEAM
THROUGH CUPOLA

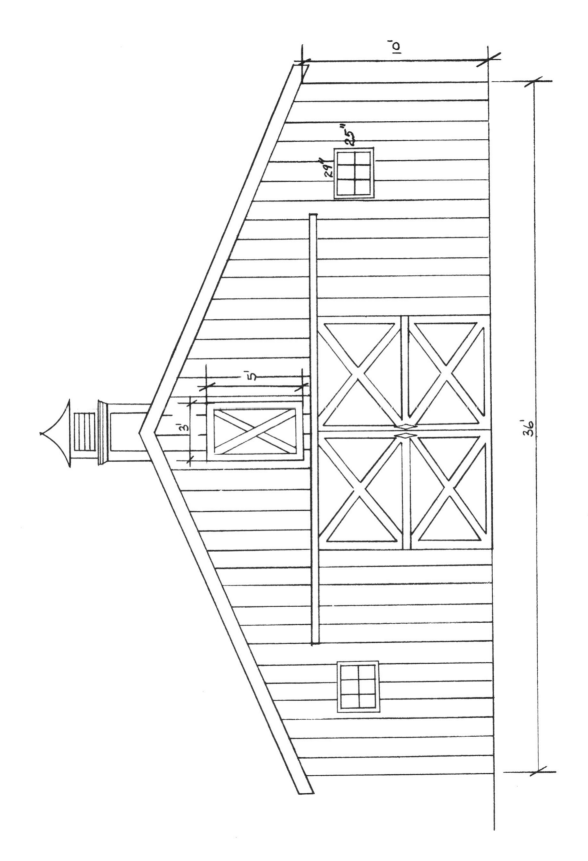

8 or 9-Horse Breeding/Training Barn

Description.

This barn, connected at one end to a spacious indoor arena, provides room for a busy collection of horses in training as well as for their groom, too. Spaces are allotted for wash stalls, a tack room, feed room and hay storage (only over the center aisle, providing good ventilation over the stalls). This design is from the barn-building company of P.J. Williams, Inc. in Somerset, Virginia.

Materials

Dimensions:
  36′ × 86′ (apartment 12′ × 24′; over-aisle loft 12′ × 86′)
Poles:
  6″ × 6″ down the aisle
  4″ × 6″ around outside of building
Framing:
  Ridge pole 2″ × 12″
  Rafter girts paired 2″ × 10″
  Loft floor girts paired 2″ × 12″
  Loft joists 2″ × 10″ 16″ o.c.
  Rafters 2″ × 8″ 4′ o.c.
  Sill girts/skirt boards 3 courses 2″ × 8″ tongue-in-groove pressure-treated lumber
Roofing:
  ½″ CDX plywood
  15-lb felt
  215-lb shingles
  2″ × 4″ purlins 2′ o.c.
Siding:
  T1-11 textured ply
Exterior doors:
  2 pair 6′ × 8′ sliders for aisle ends, 1 pair 4′ × 5′ loft doors

Windows:
  8 3′ × 4′ single-hung aluminum with bars on inside set 3″ o.c.
Trim:
  Gutters on eaves with downspouts
Interior:
  8 12′ × 12′ stalls, 1 12′ × 14′ foaling stall, 1 12′ × 12′ wash area, 1 12′ × 12′ tack room, 1 12′ × 14′ feed room, 1 12′ × 86′ over-aisle loft, 1 12′ × 24′ apartment area, plus 6′ overhang, both eaves
Stalls:
  1″ oak walls to 8′ on exterior walls, 1 ½″ oak to 4 ½′ dividers with vertical bars above 8′. Yellow pine 2″ × 6″ tongue-in-groove to 4 ½′ stall fronts with bars to 8′. 53″ × 96″ slider doors to aisle, 4′ × 8′ Dutch door to exterior
Tack room:
  Reinforced concrete floor, ½″ insulated ply walls, ½″ lauan ply ceiling, 3′ × 6′8″ solid-core lauan door with locking handle. Insulated aluminum single-hung window. Laundry tub and washer and dryer hookups installed
Wash room:
  Reinforced concrete floor with brushed finish, sloped to drain; 4″ concrete block walls with stucco finish; 1 4′ × 8′ Dutch door to exterior
Apartment:
  Reinforced concrete floor, walls paneled with 6″ insulation on exterior, fireproof sheetrock on walls adjacent to stable; ¼″ lauan ply ceiling with 6″ insulation above; 3′ × 4′ single-hung aluminum windows with screens; hollow-core lauan interior doors; exterior doors to aisle and outside 3′ × 6′8″ steel x-buck with locking handles

The stalls in this Andalusian breeding and training barn open onto a paved courtyard framed with trees. The Middleburg farm is owned by Hector Alcalde and was built by P.J. Williams, Inc. Photo by M.F. Harcourt.

The front of this barn shows room for landscaping and even a customized farm emblem on the face of the building. Photo by M.F. Harcourt.

At one end of the barn, a large indoor arena is
attached to the barn by a small breezeway.
Photo by M. F. Harcourt

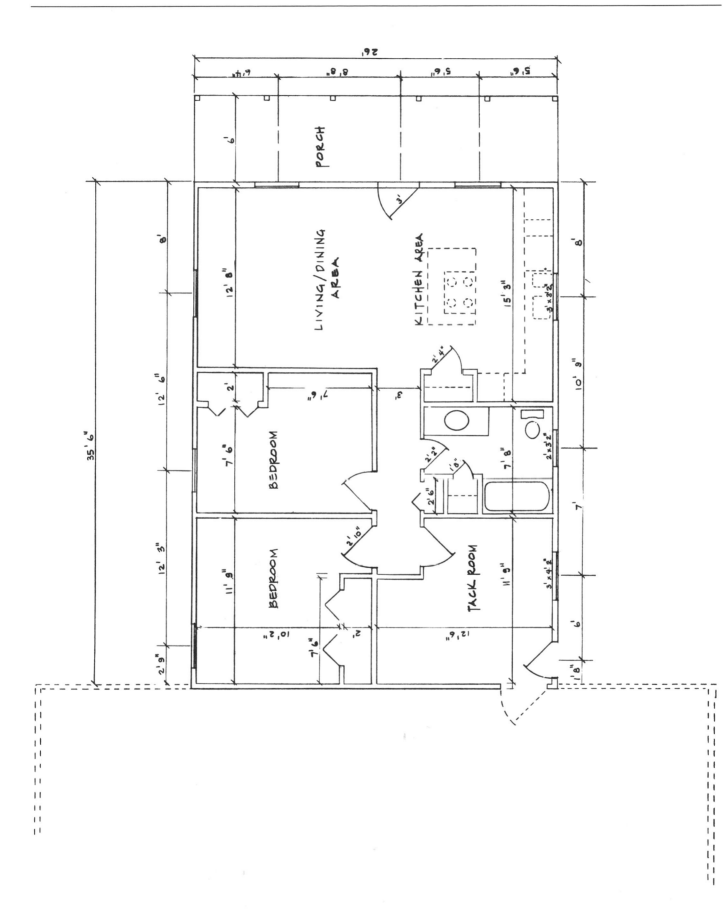

8-HORSE BARN WITH HOME OVERHEAD

Description.

This handsome home and barn combination is the design of architect David Cooper and his wife Sharon, who received advice from the Umbaugh Pole Building Company. It combines standard pole barn construction with a little of what its owners call "frou frou", such as the louvered vents at either end of the aisle doors. The vents fill the space added when the 10-foot aisle was expanded to 14 feet so that the 2,400 square foot upstairs home would not have obtrusively placed columns.

The tack room doubles as furnace room and entrance to the upstairs home. To bring light and style to this high-traffic area, the Coopers decided to use glass walls. In fact, glass is the essential ingredient to much of the home's charm, with balconies cut through the roof in three directions.

Inside the barn, they used tongue-in-groove lumber to provide a polished, neat look. A layer of fire-retardant drywall along the ceilings keep the insulation, household wiring and plumbing from the horses' reach.

Materials

Poles:
  6″ × 6″ pressure-treated
Framing:
  Sill girts double 4″ × 4″ p.t. sections
  Rafters 12″ × 2″ or Grade 2 10″ × 2″ at 24″ o.c. coupled at posts with ½″ CDX sheathing ply, joined at midpoint of column with bolted connectors.
  Purlins spaced flush with 10″ × 2″ column post
  Joists 10″ × 2″ 24″ o.c. with purlins finished with ⅝″ drywall firescreen on underside
  Interior stable framing 4″ × 4″
  Stalls lined with 2″ × 6″ tongue-in-groove oak to 42″ height, metal rods set 3″ o.c. vertically
  Deck boards 2″ × 4″ on bearers above galvanized pan
  Upstairs 2″ × 6″ stud wall framing 2′0″ o.c.
Exterior:
  T1-11 with cedar trim, louvers at either side of exterior sliders
Flooring:
  Rough-finished concrete aisle: stalls packed soil
  Outside stall area beneath overhang 12″ × 12″ pavers edged in brick
  Upstairs ¾″ ply underlayment fixed and glued, kitchen/bath/laundry tile floor, hardwood and carpets in bedrooms, living room, etc.
Doors:
  2 sliders either end
  Interior sliders each stall, with central window filled by 3″ o.c. metal rods
  Dutch doors to exterior framed with 2″ × 6″, faced with same ply sheathing as wall
  Exterior entrance to tack room and stairs area walled with fixed glass screen and fully glazed door

Can you believe this is a pole barn? It is the rear view of the beautiful house/barn combination built by David and Sharon Cooper of Aldie, Va. The basic pole barn idea was supplied by the Umbaugh Barn Company of Pennsylvania; then architect David blended all Sharon's preferences into the final equation.

The balcony, accessed by sliding glass doors, affords a view of the pastures where the Coopers keep their horses. It fits neatly over the 14-foot wide center aisle Sharon requested. Photo by N. W. Ambrosiano

This insulation detail shows how the barrier between barn below and home above is arranged. Also shown is a good, secure barn light that, while not recessed, is shielded with heavy screening to be safe from a rearing horse's head. (Cooper Farm, Aldie, Va.) Photo by N. W. Ambrosiano

The north end of this house/barn combination shows the airy window arrangement upstairs and unusual louvers downstairs that bring additional ventilation into the extra-wide barn aisle. (Cooper Farm, Aldie, Va.) Photo by N. W. Ambrosiano

The view from the southern end of the Coopers' home is maximized with a full wall of glass; both windows and sliding doors open onto a deck. (Cooper Farm, Aldie, Va.) Photo by N. W. Ambrosiano

The southern exposure of the building allows maximum sunshine into the upstairs home. Stalls, at left, are shaded from both sun and rain. (Cooper Farm, Aldie, Va.) Photo by N. W. Ambrosiano

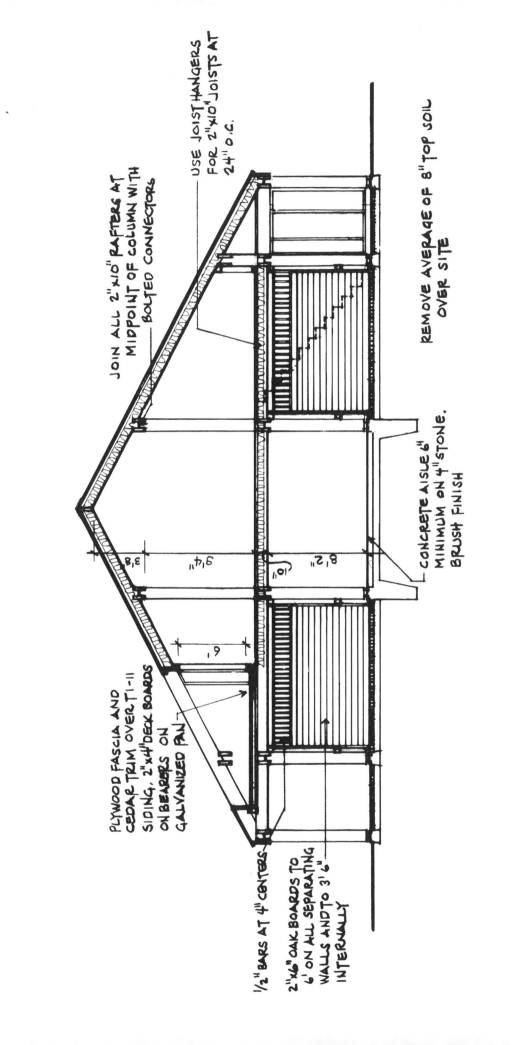

JOIN ALL 2"x10" RAFTERS AT MIDPOINT OF COLUMN WITH BOLTED CONNECTORS

USE JOIST HANGERS FOR 2"x10" JOISTS AT 24" O.C.

REMOVE AVERAGE OF 8" TOP SOIL OVER SITE

CONCRETE AISLE 6" MINIMUM ON 4" STONE. BRUSH FINISH

PLYWOOD FASCIA AND CEDAR TRIM OVER T1-11 SIDING. 2"x4" DECK BOARDS ON BEARERS ON GALVANIZED PAN

1/2" BARS AT 4" CENTERS

2"x6" OAK BOARDS TO 6' ON ALL SEPARATING WALLS AND TO 3' 6" INTERNALLY

3' 8"

9' 4"

10'

8' 2"

6'

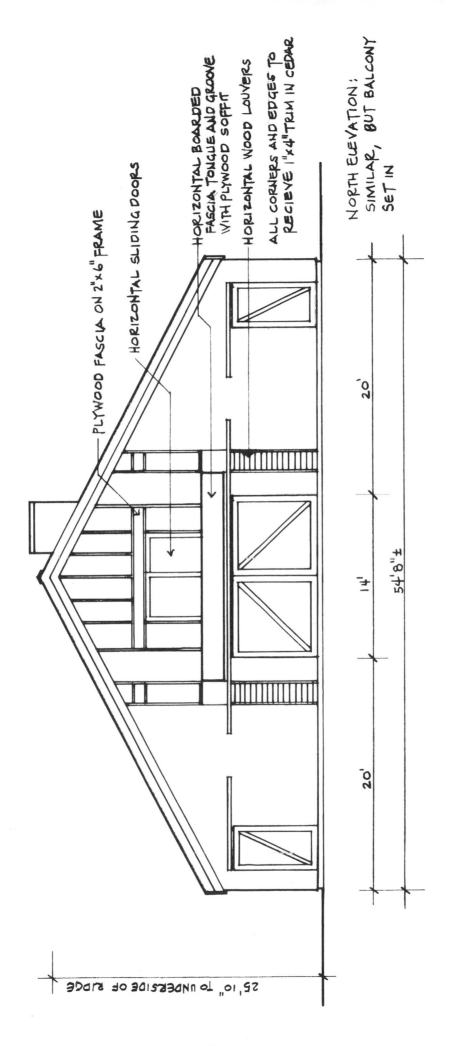

PLYWOOD FASCIA ON 2"x6" FRAME

HORIZONTAL SLIDING DOORS

HORIZONTAL BOARDED FASCIA TONGUE AND GROOVE WITH PLYWOOD SOFFIT

HORIZONTAL WOOD LOUVERS

ALL CORNERS AND EDGES TO RECIEVE 1"x4" TRIM IN CEDAR

NORTH ELEVATION; SIMILAR, BUT BALCONY SET IN

25'10" TO UNDERSIDE OF RIDGE

20'

14'

20'

54'8" ±

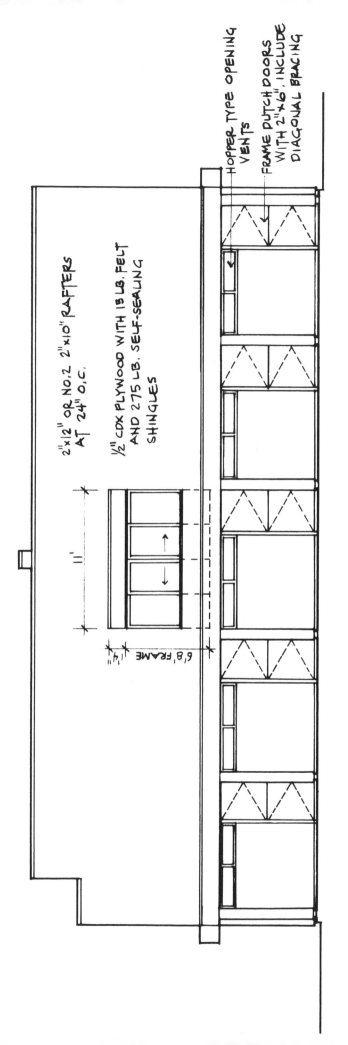

2"×12" OR NO.2 2"×10" RAFTERS AT 24" O.C.

½" CDX PLYWOOD WITH 15 LB. FELT AND 275 LB. SELF-SEALING SHINGLES

11'

6'8' FRAME

HOPPER TYPE OPENING VENTS

FRAME DUTCH DOORS WITH 2"×6" INCLUDE DIAGONAL BRACING

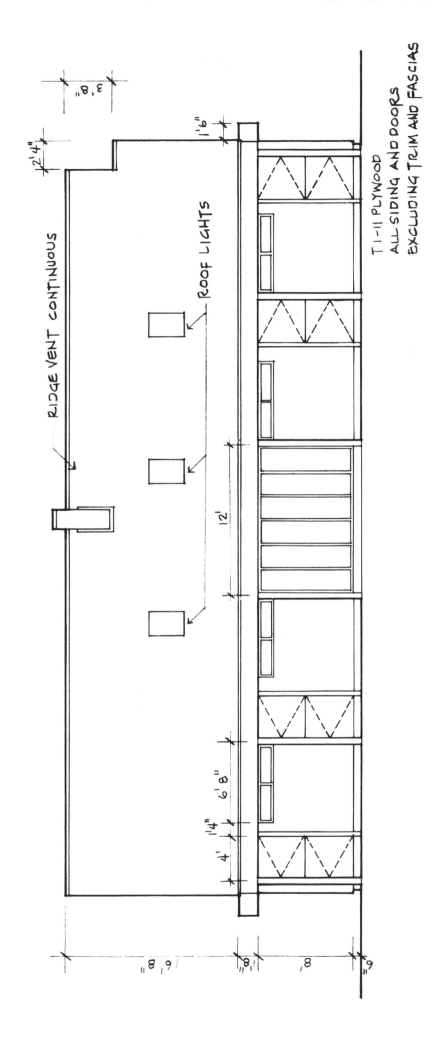

RIDGE VENT CONTINUOUS

ROOF LIGHTS

T I-II PLYWOOD
ALL SIDING AND DOORS
EXCLUDING TRIM AND FASCIAS

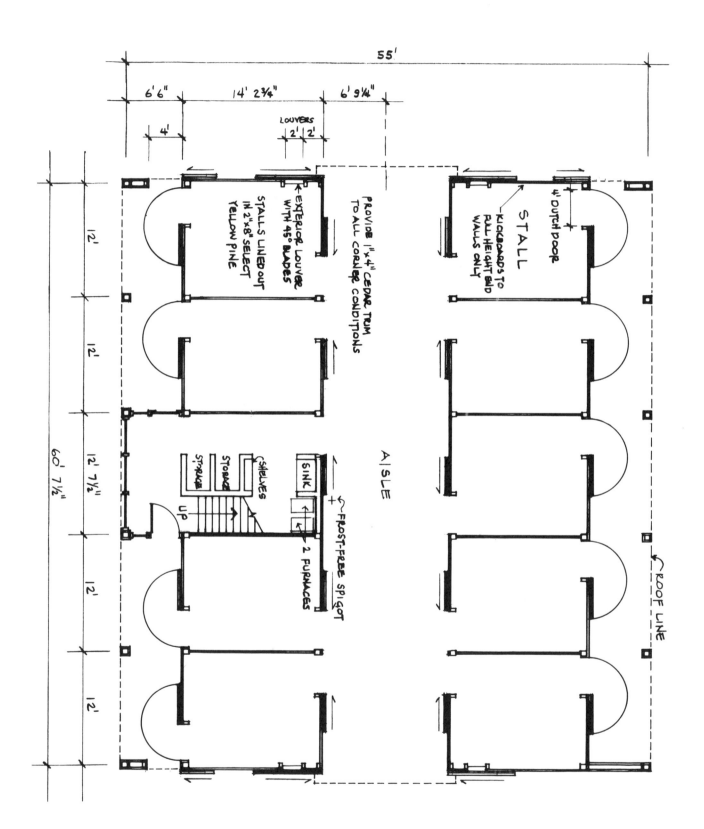

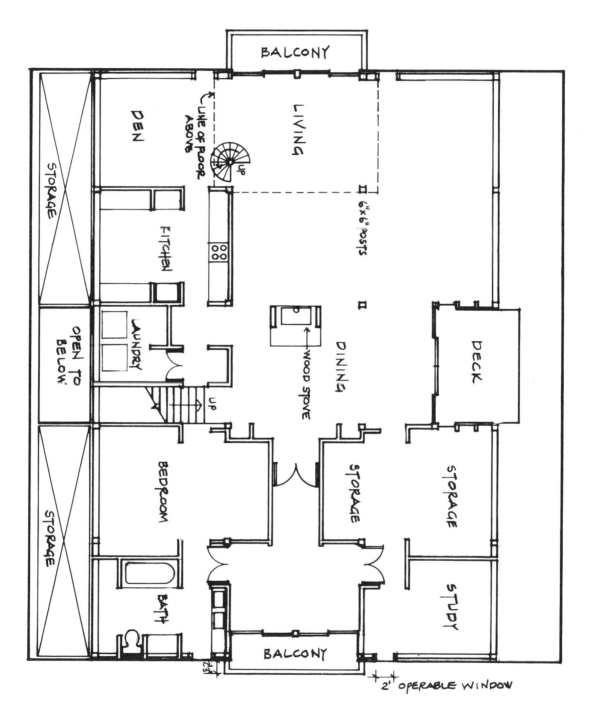

CONVERTIBLE BARN/GARAGE

Description.

This barn is ideal for a home in suburbia that needs resale flexibility. It starts as a garage with doorways facing the street and sliding doors at either end. Then, when partitions are added between the stalls and the garage-door openings are filled with the same lumber, it is transformed into a working stable.

The loft is designed to hold a full 20 tons of hay, although it can be converted to a small apartment or made with a lower pitch to the roof. This barn was built on a cinderblock foundation to overcome a very wet piece of land; it can be built as a pole structure as well. It was designed and built by Penny and Lloyd Burger of Chesapeake, Virginia.

Materials

Dimensions:
  30′ × 36′
Poles:
  6″ × 6″ inside supports on concrete footers, outer framing 4″ × 4″ and 2″ × 4″ platform-style frame on cinderblock foundation.
Framing:
  Loft floor joists 2″ × 12″ set 16″ o.c. supported by either built-up wood beam of 2″ × 12″ or steel I beam on 6″ × 6″ post with concrete footers. Garage door openings framed with 2 2″ × 4″ jacks to support header of 2 2″ × 10″s
Roofing:
  ½″ plywood deck with roofing felt and asphalt shingles to match nearby structures
Siding:
  1 × 12″ rough-cut barn siding, or T1-11 2″ × 8″× 8′ rough-cut oak in doorways, topped with vertical rods 3″ o.c.
Trim:
  1″ × 6″ starter boards at corners, 1″ × 6″ trim boards in garage openings, 1″ × 6″ fascia boards
Doors:
  2 5′ sliders at each end, 6 4′ × 8′ sliders for stalls, or screens as desired
Windows:
  3′ × 3′ openings framed with 2″ × 4″ with rods set 3″ o.c.

This barn, owned by Penny and Lloyd Burger of Chesapeake, Va., can be converted to a garage in a matter of a few hours. Built in an area of difficult drainage, it sits on a cinderblock foundation. It is faced with T1-11 textured-plywood stained to match the nearby house; the asphalt-shingle roof also matches the house. Photo by M. F. Harcourt

The convertible barn was built with 8-foot garage-door openings in the walls toward the house. Then, each was filled with 2″ × 8″ planks topped with metal bars for ventilation and light. (Burger Farm, Chesapeake, Va.) Photo by M. F. Harcourt

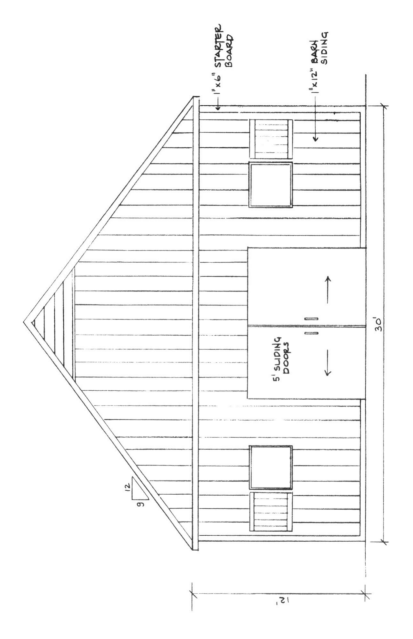

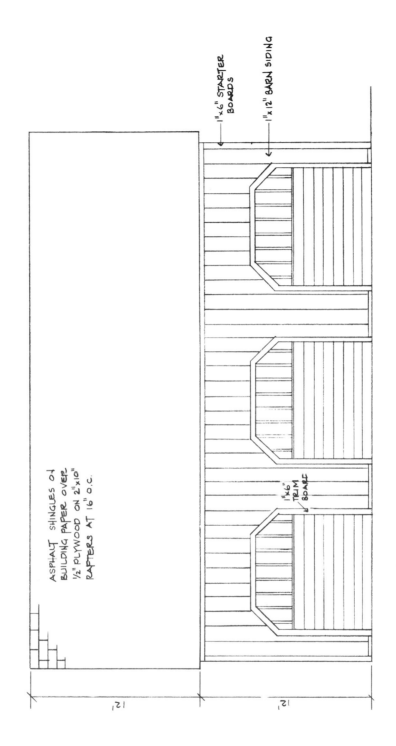

ASPHALT SHINGLES ON
BUILDING PAPER OVER
½" PLYWOOD ON 2"x10"
RAFTERS AT 16" O.C.

1"x6" STARTER BOARDS

1"x12" BARN SIDING

1"x6" TRIM BOARD

12'

12'

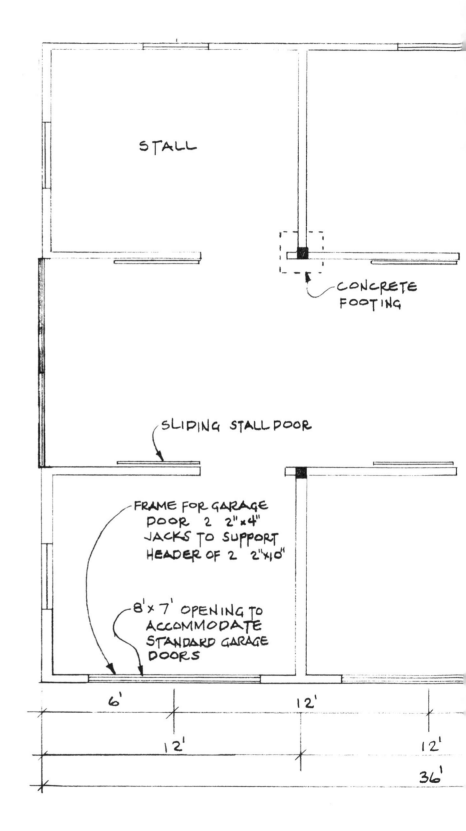

STALL

CONCRETE
FOOTING

SLIDING STALL DOOR

FRAME FOR GARAGE
DOOR 2 2"x4"
JACKS TO SUPPORT
HEADER OF 2 2"x10"

8' x 7' OPENING TO
ACCOMMODATE
STANDARD GARAGE
DOORS

6'   12'   12'

12'   12'

36'

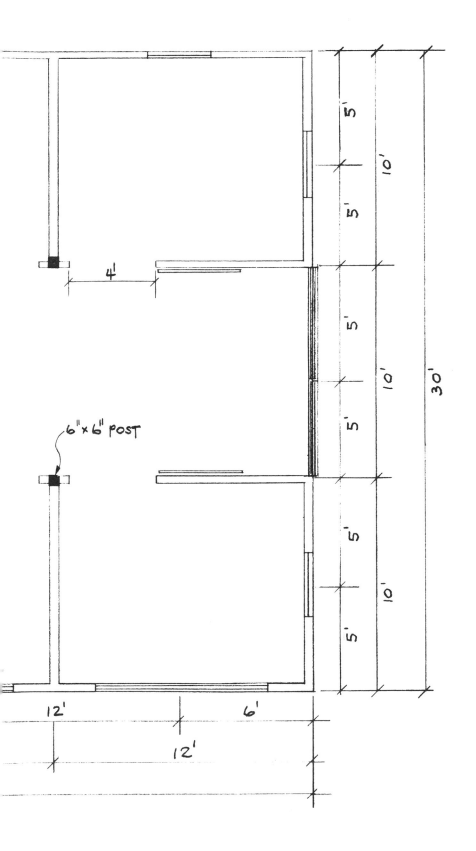

6"x 6" POST

4'

5'

5'

10'

5'

5'

10'

30'

5'

5'

10'

12'

6'

12'

HOT WEATHER BARNS

Description.

The numerous and old and new facilities located throughout the warm states, while reflecting varied climates and needs, have several features in common.

Larger aisleways or smaller stall units in each structure either remove excess heat or reduce the amount of heat by having smaller numbers of horses per area.

They contain larger-than-normal ventilation areas. Large windows, full length roof vents/cupolas, half-solid stall partitions, or wide gaps at the tops of full partitions to allow air circulation are some of the construction features that encourage ventilation. Often doors are full-length, heavy-gauge wire to allow increased circulation.

They are strategically located in relation to the prevailing winds and terrain, taking full advantage of both.

They use natural shade from trees whenever possible.

This simple pole barn represents the most versatile of buildings, actually being a wall-less collection of stalls beneath an airy roof. While shown with heavy lumber here, as might suit a draft-horse facility, the stalls can be framed in rough-cut 1″ × 6″s with 3-inch gaps between boards for additional air flow. This plan #6148 is distributed through the Marion County, Florida, Cooperative Extension Service.

Materials

Dimensions:
    40′ × 240′ (modifiable in all directions)
Poles:
    8″ diameter main supports, treated poles or telephone poles
Framing:
    Rafter girts 2″ × 8″
    Rafters 2″ × 8″ with purlins 24″ o.c.
    Stall posts 4″ × 4″ posts or 4″ diameter poles with rough-cut 2″ × 6″ or larger partitions spaced 3″
Roof:
    Corrugated metal
Details:
    Tack rooms and feed areas can be substituted for any stalls by using ½″ ply walls over 2″ × 4″ girts instead of rough-cut board partitions. Aisle and stall dimensions can be expanded as desired, since roof weight is light and supporting poles/rafters are sufficient to accept wider span.

This tile-roofed California breeding farm makes use of the long growing season, spreading cool ivy around the overhang, but containing the growth around the posts in horse-proof wire mesh. The breeze beneath is cool and soft, no matter the weather. (Nicasio Arabians, Nicasio, Calif.) Photo by N. W. Ambrosiano

This California barn features the mission-style tile roof, stucco walls and rounded windows typical in the housing of the Southwest and West. Inside, open rafters above the stalls and aisle doors that open in four directions capture every passing breeze. (Windfield Station, Nicasio, Calif.) Photo by N. W. Ambrosiano

An alternate structure for hot weather climate is constructed of cinderblock. While expensive to construct, mainly due to labor costs, the cinderblock barn, as shown, can be a cool one. This structure incorporates all the best that natural ventilation can offer. Large trees offer natural shade. In addition, the lay of the facility takes full advantage of a natural draw between the hills, channeling any breezes or natural air flow down through the center aisle. The center aisle is exceptionally wide, as are the doors and windows. In the interior, the upper area has been left totally open, with vents at each end. Doors and walls were constructed with bars on the upper half to increase airflow.

Many safety features have been incorporated into this barn, such as rounding all exposed cinderblock corners to reduce nicked horse hips. Since this is a training facility, with frequent shipments of young stock to be broken for the track, this barn is constantly being thoroughly cleaned in preparation for a new crop of highly susceptible youngsters. (September Farm, Ocala, Fla.)

The importance of sturdy interior construction is obvious when you examine the residents in their stalls. (Briar Patch Farm, Micapony, Fla.) Photo by M. F. Harcourt

A variation on the cinderblock structure allows the stalls to be backed up to each other with the aisleway fronting them on the outside. Ventilation has been increased by adding full heavy-gauge wire doors along with wide ventilation strips on the three inside walls. Since this is a breeding facility, stall separations are full, and air flow is maintained by the addition of venting blocks spaced strategically in the walls. Full use of natural shade from the trees also keep this facility cool. (Hassell Arabian Stud, Reddick, Fla.) Photo by M. F. Harcourt

This variation on the standard open stall makes use of tie stalls. This allows more animals to be handled in the same area, which is ideal for horses used constantly throughout the day. They can then stand quietly between sessions. These stalls are not recommended for a horse's full-time shelter, however, as they are far too confining. (Briar Patch Farm, Micapony, Fla.) Photo by M. F. Harcourt

The aisleways in this airy pole barn can be constructed so they are wide enough for working with any equipment or animal. The feed room is the enclosed area to the left that has been placed on a concrete floor to protect the feed. The center aisle is made of asphalt for easy cleaning and to facilitate the movement of the heavy vehicles used by the owner. (Briar Patch Farm, Micapony, Fla.) Photo by M. F. Harcourt

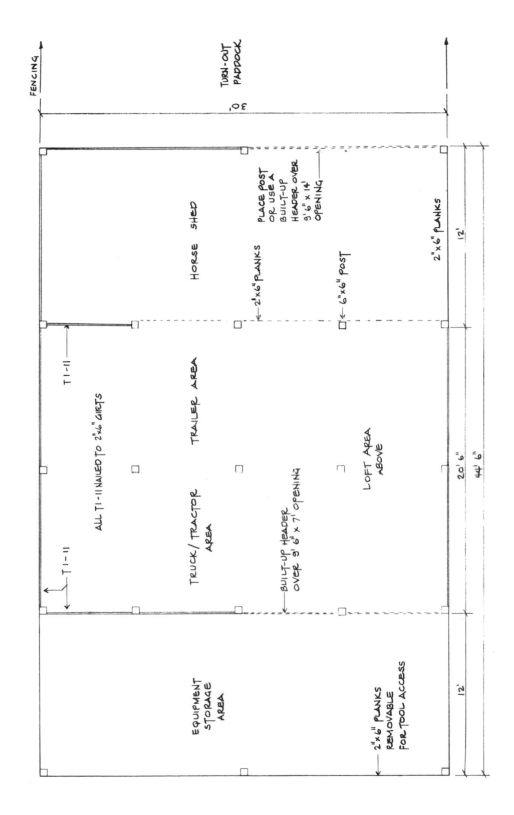

FENCING

TURN-OUT
PADDOCK

30'

HORSE SHED

PLACE POST
OR USE A
BUILT-UP
HEADER OVER
9' 6" x 14'
OPENING

2"x6" PLANKS

6"x6" POST

2"x6" PLANKS

12'

T1-11

ALL T1-11 NAILED TO 2"x6" GIRTS

TRAILER AREA

TRUCK / TRACTOR
AREA

LOFT AREA
ABOVE

BUILT-UP HEADER
OVER 9' 6" x 7' OPENING

T1-11

20' 6"

44' 6"

EQUIPMENT
STORAGE
AREA

2"x6" PLANKS
REMOVABLE
FOR TOOL ACCESS

12'

Resource List

The following are sources for plans, diagrams and materials noted in this book and elsewhere.

Agway Buildings
P.O. Box 4853
Syracuse, NY 13221
1-800-448-3400, ext. 146

American Building Systems
Division of Northern Homes
51 Glenwood Avenue
Glens Falls, NY 12801
518-798-6007

Aqua Thaw, Inc.
1208B S. Hudson St.
Tulsa, OK 74112
1-800-541-7501

Barnmaster Inc.
777 Gable Way
El Cajon, CA 92020
619-441-9400

Bonanza Buildings
(Bonanza-Umbaugh barns)
P.O. Box 9
West Route 316
Charleston, IL 61920
1-800-637-2046

Penny and Lloyd Burger
(garage/barn plans)
1060 Taft Road
Chesapeake, VA 23322
804-421-3845

Butler Rural Systems
7400 E. 13th St.
Kansas City, MO 64126
1-800-328-5860

BYB Systems Inc.
14 Hanover St.
Hanover, MA 02339
1-800-247-2767

Chore-Time Equipment Co.
(automatic feed systems)
"Triple Crown" PVC fencing
P.O. Box 518
State Rd. 15 North
Milford, IN 46542
219-658-4101

David Cooper
(8 stall barn/house combination plans)
Oldham and Seltz, Architects
21 Dupont Circle, N.W.
Washington, D.C. 20036
202-822-9797

Country Manufacturing Co.
(thermal buckets, manure spreaders, etc.)
P.O. Box 104 C-5
Fredericktown, OH 43091
614-694-9926

D.T. Industries, Inc.
97 Thames Rd.
Exeter, Ontario,
Canada, NOM 1S3
519-235-1445

Equestrian Plastics
25215 Stanford Ave.
Valencia, CA 91355
1-800-369-0303

Equine Comfort
25 Gristmill Lane
Pawling, NY 12564
914-855-5123

Equine Safety Surfaces, Inc.
P.O. Box 271256
Oklahoma City, OK 73137
405-946-7699

Equustall
9411 Burge Avenue
Richwood, VA 23237
1-800-448-3636

Farnam Companies, Inc.
P.O. Box 34820
Phoenix, AZ 85067
602-285-1660

Feed Master
331 S. River Dr., Ste. 14
Tempe, AZ 85281
602-968-0831

Fortrex/Fortiflex
(buckets)
G.P.O. Box 4523
San Juan, PR 00936 USA

Goode, Inc.
7400 Smithfield Road
P.O. Box 820793
Fort Worth, TX 76180
817-498-2299

Grain Systems, Inc.
P.O. Box 20
E. Illinois St.
Assumption, IL 62510
217-226-4421

Handi-Klasp
1519 James Street
Webster City, IA 50595
1-800-332-7900 or
515-832-5579

Humane Manufacturing Co.
805 Moore Street
Baraboo, WI 53913
1-800-248-6263

International Grating, Inc.
7625 Parkhurst
Houston, TX 77028
713-633-8614

Kraiburg Corporation
600 W. Jackson, Ste. 580
Chicago, IL 60606
312-648-8804

Linear Rubber Products, Inc.
5525 19th Avenue
Kenosha, WI 53148
1-800-558-4040,
in Wisconsin call collect:
414-652-3912

Maidware Products
200 Milliken Drive, SE
Newark Industrial Park
Hebron, OH 43025
1-800-451-3837

Midwest Plan Service
(Many versatile building
and equipment plans)
Agricultural Engineering
122 Davidson Hall
Iowa State University
Ames, IA 50011
515-294-4337

Morton Buildings
P.O. Box 399
Morton, IL 61550
1-800-447-7436

Nelson Manufacturing Co.
(automatic waterers)
3049 12th St., S.W.
Cedar Rapids, IA 52406
319-363-2607

Noland Manufacturing Co.
(weatherproof waterers)
100 East School St.
Carlisle, IA 50047
1-800-247-0037

Northwest Rubber Mats
17900 Kennedy Drive
Pitt Meadows, B.C.
Canada V3Y 1Z1
604-465-9944

Peoples Pole Buildings
29 Erie Street
Hubbard, OH 44425
216-534-1108

Petersen Company
1527 4th Avenue, S.
Dennison, IA 51442
712-263-2442

P.J. Williams Co.
(barns)
Somerset, VA 22972
703-832-2238

PortaStall
P.O. Box 1627
Mesa, AZ 85201
602-834-8812

Ritchie Industries
120 South Main
Box 780
Conrad, IA 50621
515-366-2525

Rockin J Horse Stalls
Box 896
Mannford, OK 74044
918-865-3366

Rohn Agri Products
P.O. Box 2000
Peoria, IL 61656
1-800-447-2264

Rubbermaid
3124 Valley Avenue
Winchester, VA 22601
703-667-8700

Rubbertech, Inc.
34 Water Street
Excelsior, MN 55331
612-470-0858

Steelmaster Buildings
1023 Laskin Road, Ste. 109
Virginia Beach, VA 23451
1-800-341-7007

Upperville Barns
Division of Northern Counties
P.O. Box 97
Upperville, VA 22176
703-592-3232

VaFaC
(pre-fab stalls, equipment,
ventilators, etc.)
P.O. Box 7931
Fredericksburg, VA 22404
703-898-5425

White House Trading Co.
12610 LaGrange Rd.
Louisville, KY 40245
1-800-782-5628

Woodstar Products, Inc.
(stall doors, hardware)
P.O. Box 444
Delavan, WI 53115
414-728-8460

W-W Manufacturing Co., Inc
P.O. Box 728
Dodge City, KS 67801
316-227-7111

INDEX